Sum Formula for SL₂ over a Totally Real Number Field

MEMOIRS
of the
American Mathematical Society

Number 919

Sum Formula for SL_2 over a Totally Real Number Field

Roelof W. Bruggeman
Roberto J. Miatello

January 2009 • Volume 197 • Number 919 (first of 5 numbers) • ISSN 0065-9266

American Mathematical Society
Providence, Rhode Island

2000 *Mathematics Subject Classification.* Primary 11F72, 11F30, 11F41, 11L05, 22E30.

Library of Congress Cataloging-in-Publication Data

Bruggeman, Roelof W., 1944–
  Sum formula for SL_2 over a totally real number field / Roelof W. Bruggeman and Roberto J. Miatello.
      p. cm. — (Memoirs of the American Mathematical Society, ISSN 0065-9266 ; v. 197, no. 919)
  Includes bibliographical references and index.
  ISBN 978-0-8218-4202-7 (alk. paper)
  1. Lie groups. 2. Representations of groups. I. Miatello, Roberto J. II. Title.

QA387.B75 2009
512′.482—dc22   2008039456

Memoirs of the American Mathematical Society

This journal is devoted entirely to research in pure and applied mathematics.

Subscription information. The 2009 subscription begins with volume 197 and consists of six mailings, each containing one or more numbers. Subscription prices for 2009 are US$709 list, US$567 institutional member. A late charge of 10% of the subscription price will be imposed on orders received from nonmembers after January 1 of the subscription year. Subscribers outside the United States and India must pay a postage surcharge of US$65; subscribers in India must pay a postage surcharge of US$95. Expedited delivery to destinations in North America US$57; elsewhere US$160. Each number may be ordered separately; *please specify number* when ordering an individual number. For prices and titles of recently released numbers, see the New Publications sections of the *Notices of the American Mathematical Society*.

Back number information. For back issues see the *AMS Catalog of Publications*.

Subscriptions and orders should be addressed to the American Mathematical Society, P. O. Box 845904, Boston, MA 02284-5904, USA. *All orders must be accompanied by payment.* Other correspondence should be addressed to 201 Charles Street, Providence, RI 02904-2294, USA.

Copying and reprinting. Individual readers of this publication, and nonprofit libraries acting for them, are permitted to make fair use of the material, such as to copy a chapter for use in teaching or research. Permission is granted to quote brief passages from this publication in reviews, provided the customary acknowledgment of the source is given.

Republication, systematic copying, or multiple reproduction of any material in this publication is permitted only under license from the American Mathematical Society. Requests for such permission should be addressed to the Acquisitions Department, American Mathematical Society, 201 Charles Street, Providence, Rhode Island 02904-2294, USA. Requests can also be made by e-mail to `reprint-permission@ams.org`.

Memoirs of the American Mathematical Society (ISSN 0065-9266) is published bimonthly (each volume consisting usually of more than one number) by the American Mathematical Society at 201 Charles Street, Providence, RI 02904-2294, USA. Periodicals postage paid at Providence, RI. Postmaster: Send address changes to Memoirs, American Mathematical Society, 201 Charles Street, Providence, RI 02904-2294, USA.

© 2009 by the American Mathematical Society. All rights reserved.
Copyright of individual articles may revert to the public domain 28 years
after publication. Contact the AMS for copyright status of individual articles.
This publication is indexed in *Science Citation Index*®, *SciSearch*®, *Research Alert*®,
CompuMath Citation Index®, *Current Contents*®/*Physical, Chemical & Earth Sciences*.
Printed in the United States of America.

∞ The paper used in this book is acid-free and falls within the guidelines
established to ensure permanence and durability.
Visit the AMS home page at `http://www.ams.org/`

10 9 8 7 6 5 4 3 2 1   14 13 12 11 10 09

Contents

Introduction 1

Chapter 1. Spectral sum formula 3
   1.   Sum formulas of Kuznetsov type 3
   2.   Preliminaries 5
   3.   Derivation of the spectral sum formula 16
   4.   Density results for cuspidal representations 46

Chapter 2. Kloosterman sum formula 55
   5.   Bessel inversion 55
   6.   Derivation of the Kloosterman sum formula 60
   7.   Application 68
   8.   Final comments 70

Appendix A. Sum formula for the congruence subgroup $\Gamma_1(I)$ 71

Appendix B. Comparisons 73

Bibliography 77

Index 79

Abstract

We prove a general form of the sum formula for SL_2 over a totally real number field. This formula relates sums of Kloosterman sums to products of Fourier coefficients of automorphic representations. We give two versions: the spectral sum formula (in short: sum formula) and the Kloosterman sum formula. They have the independent test function in the spectral term, in the sum of Kloosterman sums, respectively.

The discrete subgroup is $\Gamma_0(I)$, where I is a non-zero ideal in the integers of the number field. We allow a character of the form $\begin{pmatrix} a & b \\ c & d \end{pmatrix} \mapsto \chi(d)$, with χ a character modulo I.

As an application, we obtain density results for cuspidal representations, extending results in [9].

Received by the editor Received by editor July 26, 2005, and in revised form April 12, 2006.
2000 *Mathematics Subject Classification*. Primary 11F72 11F30 11F41 11L05; Secondary 22E30.
Key words and phrases. Automorphic representation, Bessel transformation, Kloosterman sum, sum formula.
Roberto J. Miatello partially supported by Conicet and by grants from Foncyt, AgCba, SecytUNC and Fund. Antorchas.

Introduction

Kloosterman sums arose as a tool in the study of quadratic forms, see Kloosterman, [**26**]. The relation with Fourier coefficients of holomorphic modular cusp forms became explicit in [**27**]. Petersson expressed Fourier coefficients of Poincaré series as series with Kloosterman sums and Bessel functions, see equation (g) on p. 178 of [**40**]. The estimate of Weil of individual Kloosterman sums, see [**52**], was used by Selberg to derive information concerning the automorphic spectrum, see [**43**].

Petersson's formula for the Fourier coefficients of Poincaré series is restricted to individual automorphic forms. To apply it in the context of Maass forms, one should look at many automorphic forms at the same time. Bruggeman, [**3**], obtained a sum formula and used it to derive density results for the spectrum of Maass forms for $SL_2(\mathbb{Z})$. Kuznetsov, [**29**], gave independently a more versatile sum formula, with which one can detect cancellation between Kloosterman sums.

The purpose of this paper is to state and prove a general form of the sum formula for SL_2 over a totally real number field. The discrete subgroup is the congruence subgroup $\Gamma_0(I)$, where I is a non-zero ideal in the ring of integers. By allowing a character of the form $\begin{pmatrix} a & b \\ c & d \end{pmatrix} \mapsto \chi(d)$, with χ a character modulo I, we effectively deal with $\Gamma_1(I)$. Any congruence subgroup contains a conjugate of some $\Gamma_1(I)$.

The first main result is Theorem 3.21, the spectral formula, stated in §3.6 with a detailed explanation of the terms occurring in it. This formula is a tool to obtain results concerning spectral data, like those in Theorem 4.1. The other main result is Theorem 6.5, the Kloosterman sum formula. It is obtained from Theorem 3.21 by inversion of a Bessel transformation. To extend it to a wide class of test functions, precise estimates of the Bessel transforms are studied. The Kloosterman sum formula can be used to derive results concerning Kloosterman sums. These sum formulas concern automorphic forms on $\Gamma_0(I)$. In the appendix, we give the corresponding sum formulas for $\Gamma_1(I)$.

The proof follows the approach in [**7**], using the integral transformations in [**4**] and ideas from [**35**] and [**10**]. New aspects are the use of SL_2 instead of PSL_2, and the enlargement of the classes of test functions as far as the present methods allow. Also new is the explicit application to products of $SL_2(\mathbb{R})$ of the extension method in §3.5. This approach was used for $SL_2(\mathbb{C})$ in [**10**] and [**33**]. New are also the use of multiplicative characters and the application to $\Gamma_1(I)$.

We bring together methods and ideas from various publications, and have tried to explain carefully the many technicalities involved in carrying out the proofs. We hope that this work will enhance the accessibility of the sum formula, and lead to further applications. In §4, we work out one application, on the density of cuspidal representations having local components of a prescribed type, extending results in [**9**].

Also, as mentioned in §7, Theorem 6.5 can be applied to yield estimates of sums of Kloosterman sums, like in [**7**] and [**8**]. We do not completely carry out the calculations in this paper, but we expect that similar estimates as in [**7**] can be reached, showing cancellation between Kloosterman sums.

We thank the referee for a careful reading of the paper and several useful suggestions that have improved the exposition.

CHAPTER 1

Spectral sum formula

1. Sum formulas of Kuznetsov type

First we discuss the sum formula in the context of $\mathrm{SL}_2(\mathbb{Z})\backslash\mathrm{SL}_2(\mathbb{R})$. We start with Theorem (3.29) in [**3**]. This result concerns cuspidal Maass forms on the upper half plane $\mathfrak{H}$. These are functions u on $\mathfrak{H}$ that satisfy $u\left(\frac{az+b}{cz+d}\right) = u(z)$ for all $\begin{pmatrix} a & b \\ c & d \end{pmatrix} \in \mathrm{SL}_2(\mathbb{Z})$ and have a Fourier expansion

$$(1.1) \qquad u(x+iy) = \sum_{n\neq 0} \gamma(n) e^{2\pi i n x} W_{0,\nu}(4\pi|n|y),$$

with $\nu \in i\mathbb{R}$. The Whittaker function $W_{0,\nu}$ can be expressed in terms of the K-Bessel function: $W_{0,\nu}(t) = \sqrt{t/\pi} K_\nu(t/2)$.

It is known that there is a countable orthonormal system $\{u_j : j \geq 1\}$ spanning the cuspidal part of $L^2(\mathrm{SL}_2(\mathbb{Z})\backslash\mathfrak{H})$, with spectral parameters $\nu_j \in i(0,\infty)$. A normalization can be made such that each u_j is a simultaneous Hecke eigenvector:

$$(1.2) \qquad \sum_{d|n} \sum_{b \bmod d} u_j\left(\frac{b+nz/d}{d}\right) = \lambda_j(n) u_j(z) \qquad (n \geq 1)$$

$$(1.3) \qquad u_j(-\bar{z}) = \varepsilon_j u_j(z).$$

Then $c_j(1) \neq 0$. Requiring $c_j(1) > 0$ fixes the normalization.

The Eisenstein series $E_\nu(z)$ is a family of $\mathrm{SL}_2(\mathbb{Z})$-invariant functions on $\mathfrak{H}$ that are not square integrable. These functions have the Fourier expansion

$$(1.4) \qquad E_\nu(x+iy) = y^{1/2+\nu} + c_{0,0}(\nu) y^{1/2-\nu}$$
$$+ \sum_{n\neq 0} c_{0,|n|}(\nu) e^{2\pi i n x} W_{0,\nu}(4\pi|n|y),$$
$$c_{0,0}(\nu) = \pi^{\nu-1/2}\Gamma(1/2-\nu)\zeta(1-2\nu)/\left(\pi^{-\nu-1/2}\Gamma(1/2+\nu)\zeta(1+2\nu)\right),$$
$$c_{0,n}(\nu) = n^{-1/2-\nu}\sigma_{2\nu}(n)/\left(\pi^{-\nu-1/2}\Gamma(1/2+\nu)\zeta(1+2\nu)\right) \qquad (n \geq 1),$$
$$\sigma_a(n) = \sum_{d|n} d^a.$$

Let the function f be holomorphic and even on some strip $|\mathrm{Re}\,\nu| \leq \tau$ with $\tau > \frac{1}{4}$, and suppose that it satisfies

$$(1.5) \qquad f(\nu) \ll (1+|\nu|)^{-2-\varepsilon} e^{-\pi|\mathrm{Im}\,\nu|}$$

for some $\varepsilon > 0$. Let n and m be positive integers. The sum formula states that for such *test functions* the following equality holds, with absolute convergence of all sums and integrals in the various terms:

$$\text{(1.6)} \quad \sum_{j\geq 1} f(v_j)\gamma_j(m)\gamma_j(n) + \frac{1}{4\pi i}\int_{\operatorname{Re} v=0} f(v)c_{0,0}(-v)c_{0,m}(v)c_{0,n}(v)\,dv$$

$$= -\frac{\delta_{m,n}}{4\pi m}\frac{1}{2\pi i}\int_{\operatorname{Re} v=0} f(v)\,2v\,\sin\pi v\,dv$$

$$+ \frac{1}{2\sqrt{nm}}\sum_{c=1}^{\infty}\frac{S(m,n;c)}{c}\tilde{f}\left(\frac{4\pi\sqrt{nm}}{c}\right).$$

The left hand side of the formula is built with products of the Fourier coefficients of the cuspidal Maass forms u_j and of the Eisenstein series E_v. The two terms correspond to the discrete and continuous spectrum of the Laplace operator on $\operatorname{SL}_2(\mathbb{Z})\backslash\mathfrak{H}$. We call this the *spectral side*.

The right hand side is related to a sum over elements $\begin{pmatrix} a & b \\ c & d \end{pmatrix}$ representing cosets in $\left\{\begin{pmatrix} 1 & * \\ & 1 \end{pmatrix}\right\}\backslash\operatorname{SL}_2(\mathbb{Z})$; we call it the *geometric side*. The first term, the *delta term*, comes from the matrices with $c=0$. It is non-zero only if $m=n$. The matrices with $c\neq 0$ give rise to the Kloosterman term. Here, we see the *Kloosterman sums*

$$\text{(1.7)} \quad S(m,n;c) = \sum_{d\bmod c}^{*} e^{2\pi i(ma+nd)/c} \qquad (ad\equiv 1\bmod c),$$

and the *Bessel transform*

$$\text{(1.8)} \quad \tilde{f}(t) = \frac{1}{2\pi i}\int_{\operatorname{Re} v=0} f(v)J_{2v}(t)\,2v\,dv.$$

The sum formula can be viewed as a generalization of Petersson's formula for the Fourier coefficients of holomorphic Poincaré series (see [40]). In the context of Maass forms, it is not possible to find in this way information concerning Fourier coefficients of an individual automorphic form. However, statistical information is possible. For example, Proposition 4.1 of [3] derives from the sum formula the following distribution result:

$$\text{(1.9)} \quad \sum_{j\geq 1} e^{-v(\frac{1}{4}-v_j^2)}\frac{\gamma_j(n)^2}{\cosh(\pi\operatorname{Im} v_j)} = \frac{1}{4\pi nv} + O\left(v^{-1/2-\varepsilon}\right) \qquad (v\downarrow 0).$$

This is obtained by a special choice of the test function f. The main contribution $\frac{1}{4\pi nv}$ is produced by the delta term.

Kuznetsov gives a similar sum formula in his preprint [28], published in [29]. Moreover, he gives a right inverse of the Bessel transform in (1.8). Thus, it is possible to use the function $\tilde{f}$ as the independent test function, which makes the sum formula more powerful. Sums of Kloosterman sums can be estimated. Kuznetsov proves that

$$\text{(1.10)} \quad \sum_{c=1}^{X}\frac{S(m,n;c)}{c} \ll_{n,m} X^{1/6}\log^{1/3} X.$$

In §5 of [3], a variant of the sum formula is given, where also the holomorphic cusp forms for $\operatorname{SL}_2(\mathbb{Z})$ are present in the spectral side. This suggests that the essential ingredients in the left hand side of (1.6) are not the cusp forms in $L^2(\operatorname{SL}_2(\mathbb{Z})\backslash\mathfrak{H})$, but the cuspidal representations of $\operatorname{SL}_2(\mathbb{R})$ in the Hilbert space $L^2(\operatorname{SL}_2(\mathbb{Z})\backslash\operatorname{SL}_2(\mathbb{R}))$. This point of view is taken up in [4].

The goal of the present paper is to derive the sum formula in the Hilbert modular case, for $\Gamma_0(I)\backslash(\operatorname{SL}_2(\mathbb{R}))^d$. The discrete group $\Gamma_0(I)$ is a Hecke congruence subgroup of the Hilbert modular group associated to a totally real number field of degree d over $\mathbb{Q}$. We allow characters of $\Gamma_0(I)/\Gamma_1(I)$. The general approach is that in [7], where we work effectively on PSL_2 (even weights), and depends on ideas in [29], [4] and [35]. The presence of a non-trivial character brings changes in the delta term and in the description of the Bessel transform. To work with test functions on a narrow strip, we use the Weil type estimate (2.47) and the fact that the norms of matrix elements of integers in the number field are rational integers.

2. Preliminaries

2.1. Hilbert modular group. We fix a totally real number field F, and denote by $\sigma_1, \ldots, \sigma_d$ the different embeddings $F \to \mathbb{R}$. We view F as a subset of $\mathbb{R}^d$ via $\xi \mapsto (\xi^{\sigma_1}, \ldots, \xi^{\sigma_d})$, and correspondingly $\mathrm{SL}_2(F)$ as a subset of $G := \mathrm{SL}_2(\mathbb{R})^d$. Under the embedding $F \subset \mathbb{R}^d$, the ring of integers O of F forms a lattice in $\mathbb{R}^d$. The image of $\mathrm{SL}_2(O)$ in G is a discrete subgroup of G with finite covolume. It is called the *Hilbert modular group*, see [14], §3.

We can place this in the framework of arithmetic groups as discussed in, e.g., §7.11 of [1]. Applying to SL_2 the functor $R_{F/\mathbb{Q}}$ of "restriction of scalars", one obtains an algebraic group $\mathbf{G}$ over $\mathbb{Q}$; see the discussion in [1], §7.16, or in [36], p. 113–121. It satisfies $\mathbf{G}_{\mathbb{Q}} \cong \mathrm{SL}_2(F)$, with $\mathbf{G}_{\mathbb{Z}} \cong \mathrm{SL}_2(O)$ as an arithmetic subgroup, and $\mathbf{G}_{\mathbb{R}} \cong \mathrm{SL}_2(\mathbb{R})^d = G$.

2.1.1. *Discrete subgroup.* We shall work with the discrete subgroup

$$(2.1) \qquad \Gamma_0(I) = \left\{ \begin{pmatrix} a & b \\ c & d \end{pmatrix} \in \mathrm{SL}_2(O) \; : \; c \in I \right\}$$

for some non-zero ideal $I \subset O$. The homomorphism

$$\mathrm{SL}_2(O) \longrightarrow \mathrm{SL}_2(O/I)$$

is surjective; see Hurwitz [20] (Satz on p. 249 of Math. Werke). The group $\Gamma_0(I)$ is the full preimage of the subgroup $\left\{ \begin{pmatrix} * & * \\ 0 & * \end{pmatrix} \in \mathrm{SL}_2(O/I) \right\}$ under this surjection. The normal subgroup

$$\Gamma_1(I) = \left\{ \begin{pmatrix} a & b \\ c & d \end{pmatrix} \in \Gamma_0(I) \; : \; a \equiv d \equiv 1 \mod I \right\}$$

of $\Gamma_0(I)$ is the full preimage of $\left\{ \begin{pmatrix} 1 & * \\ 0 & 1 \end{pmatrix} \in \mathrm{SL}_2(O/I) \right\}$ under this surjection. So $\begin{pmatrix} a & b \\ c & d \end{pmatrix} \mapsto d$ induces an isomorphism $\Gamma_0(I)/\Gamma_1(I) \longrightarrow (O/I)^*$.

We fix a character χ of $(O/I)^*$, and denote by χ also the character of $\Gamma_0(I)$ given by $\begin{pmatrix} a & b \\ c & d \end{pmatrix} \mapsto \chi(d)$. We will consider functions on G with the following automorphic transformation rule:

$$(2.2) \qquad f(\gamma g) = \chi(d) f(g) \qquad \text{for all } \gamma = \begin{pmatrix} a & b \\ c & d \end{pmatrix} \in \Gamma_0(I).$$

As $\Gamma_1(I)$ is the intersection of the kernels of the characters $\begin{pmatrix} a & b \\ c & d \end{pmatrix} \mapsto \chi(d)$, any $\Gamma_1(I)$-invariant function is a finite sum of functions satisfying (2.2), with χ running through the characters of $(O/I)^*$.

In the sequel, we shall write Γ instead of $\Gamma_0(I)$.

2.1.2. *Conventions and normalizations.* We use the notations $\mathbb{N} = \mathbb{Z}_{\geq 1}$, $\mathbb{N}_0 = \mathbb{Z}_{\geq 0}$. We often write ξ_j instead of ξ^{σ_j} for $\xi \in F$.

For $x, y \in \mathbb{R}^d$ we define $xy \in \mathbb{R}^d$ and $|x| \in [0, \infty)^d$ by $(xy)_j = x_j y_j$ and $|x|_j = |x_j|$. If $y \in (0, \infty)^d$ and $w \in \mathbb{C}$, then $y^w := \left(y_1^w, \ldots, y_d^w \right)$. The *trace* $\mathrm{Tr}_{F/\mathbb{Q}} : F \to \mathbb{Q}$ is extended to $S : \mathbb{R}^d \to \mathbb{R}$, $S(x) := \sum_{j=1}^d x_j$. Similarly, $N(x) := \prod_{j=1}^d x_j$ extends the *norm* $N_{F/\mathbb{Q}}$.

For $x, \vartheta \in \mathbb{R}, y > 0$:

(2.3) $$n(x) := \begin{pmatrix} 1 & x \\ 0 & 1 \end{pmatrix}, \quad k(\vartheta) := \begin{pmatrix} \cos\vartheta & \sin\vartheta \\ -\sin\vartheta & \cos\vartheta \end{pmatrix}, \quad a(y) := \begin{pmatrix} \sqrt{y} & 0 \\ 0 & 1/\sqrt{y} \end{pmatrix}$$

in $SL_2(\mathbb{R})$. If $x \in \mathbb{R}^d$, we put $n(x) := (n(x_1), \ldots, n(x_d)) \in G$, and similarly for $a(y)$, $k(\vartheta)$ if $y \in (0, \infty)^d$, $\vartheta \in \mathbb{R}^d$.

The group $K := \{k(\vartheta) : \vartheta \in \mathbb{R}^d\} = SO_2(\mathbb{R})^d$ is a maximal compact subgroup of G. We fix also the unipotent subgroup $N := \{n(x) : x \in \mathbb{R}^d\} \subset G$, and the group $A := \{a(y) : y \in (0, \infty)^d\}$. This group A is the connected component of 1 in the maximal $\mathbb{R}$-split torus $\{\begin{pmatrix} t & 0 \\ 0 & 1/t \end{pmatrix} : t \in (\mathbb{R}^*)^d\}$. This torus is the direct product of A and the *center* $M := \{k(\vartheta) : \vartheta \in \{0, \pi\}^d\}$ of G. It is convenient to define for $t \in (\mathbb{R}^*)^d$:

(2.4) $$m(t) := k(\pi\zeta), \quad \zeta \in \{0, 1\}^d, \quad \zeta_j = \frac{1 - \operatorname{sign} t_j}{2}.$$

In this way we have $\begin{pmatrix} t & 0 \\ 0 & 1/t \end{pmatrix} = a(t^2)m(t)$.

The *Iwasawa decomposition* $G = NAK$ gives a unique way of writing $g \in G$ as $g = nak$ with $n \in N$, $a \in A$, $k \in K$. We call (x, y, ϑ) the *Iwasawa coordinates* of $g = n(x)a(y)k(\vartheta) \in G$, and will use them often, sometimes without explicitly mentioning this.

We fix *Haar measures* on these subgroups: $dk := \frac{d\vartheta_1}{2\pi} \cdots \frac{d\vartheta_d}{2\pi}$ for $k = k(\vartheta)$ gives K volume 1. We use $dn := dx_1 \cdots dx_d$ for $n = n(x) \in N$ and $da := \frac{dy_1}{y_1} \cdots \frac{dy_d}{y_d}$ for $a = a(y) \in A$. Note that the normalization of dn differs from that in [7]. On G we use $dg := |a|^{-1} dn\, da\, dk$, with $|a(y)| := N(y)$.

The map $n(x)a(y) \mapsto (x_1 + iy_1, \ldots, x_d + iy_d)$ identifies NA with $\mathfrak{H}^d$, where $\mathfrak{H}$ is the upper half plane. The measure $dn\, da$ corresponds to the standard invariant measure

(2.5) $$d\mu(z) := \frac{dx_1\, dy_1}{y_1^2} \cdots \frac{dx_d\, dy_d}{y_d^2}$$

on $\mathfrak{H}^d$.

2.1.3. **Weights.** A function $f : G \to \mathbb{C}$ has *weight* $q \in \mathbb{Z}^d$ if it transforms on the right under K according to the character $\phi_q : k(\vartheta) \mapsto e^{iS(q\vartheta)}$ of K:

(2.6) $$f(gk(\vartheta)) = f(g) \prod_{j=1}^{d} e^{iq_j\vartheta_j}.$$

For a non-zero function f, conditions (2.2) and (2.6) can be satisfied at the same time only under the condition

(2.7) $$\chi(-1) = e^{\pi i S(q)},$$

which we impose from now on.

A function on G is called *K-finite* if it is a finite sum of functions satisfying (2.6), with q depending on the summand.

2.1.4. **Cusps.** From [14], Corollary 3.5_1, it follows that $\Gamma = \Gamma_0(I)$ has a finite number of cusp classes. Let $\mathcal{P}$ be a set of representatives of those classes. For each $\kappa \in \mathcal{P}$ we fix $g_\kappa \in \mathbf{G}_\mathbb{Q}$ such that $\kappa = g_\kappa \cdot \infty$. For the class of ∞ we choose ∞ as the representative and $g_\infty = 1$.

For each $\kappa \in \mathcal{P}$, we have the subgroup $\Gamma_\kappa := \{\gamma \in \Gamma : \gamma\kappa = \kappa\}$ of Γ fixing κ. It is contained in $g_\kappa NAMg_\kappa^{-1}$. If $d > 1$, the unipotent subgroup $g_\kappa Ng_\kappa^{-1} \cap \Gamma$ has infinite index in Γ_κ. By $\mathcal{P}_\chi$ we denote the set of $\kappa \in \mathcal{P}$ for which $\begin{pmatrix} a & b \\ c & d \end{pmatrix} \mapsto \chi(d)$ is trivial on $g_\kappa Ng_\kappa^{-1} \cap \Gamma$.

2. PRELIMINARIES

We work with the cusp ∞ as much as possible, and write Γ_N for $\Gamma \cap N$. We have

$$(2.8) \quad \Gamma_N = \left\{ \begin{pmatrix} 1 & \xi \\ 0 & 1 \end{pmatrix} : \xi \in O \right\}, \quad \Gamma_\infty = \left\{ \begin{pmatrix} \varepsilon^{-1} & \xi \\ 0 & \varepsilon \end{pmatrix} : \xi \in O, \varepsilon \in O^* \right\},$$

and $\infty \in \mathcal{P}_\chi$.

2.2. Automorphic forms. Let $q \in \mathbb{Z}^d$ satisfy (2.7). A χ-*automorphic form* of *weight* q and *spectral parameter* $\nu \in \mathbb{C}^d$ is a function $f : G \to \mathbb{C}$ satisfying the following conditions:

(A1) f satisfies the automorphic transformation rule $f(\gamma g) = \chi(d) f(g)$ for all $\gamma = \begin{pmatrix} a & b \\ c & d \end{pmatrix} \in \Gamma$.

(A2) $f(g k(\vartheta)) = f(g) e^{i S(q \vartheta)}$.

(A3) $C_j f = \left(\frac{1}{4} - \nu_j^2 \right) f$ for $j = 1, \ldots, d$ with $\nu_j \in \mathbb{C}$. By C_j we denote the Casimir operator in the j-th factor of G. In Iwasawa coordinates: $C_j = -y_j^2 \partial_{x_j}^2 - y_j^2 \partial_{y_j}^2 + y_j \partial_{x_j} \partial_{\vartheta_j}$.

It is often convenient to use the spectral parameter $\nu = (\nu_1, \ldots, \nu_d)$. It determines the vector $\left(\frac{1}{4} - \nu_j^2 \right)_j$ of eigenvalues of the Casimir operators.

(A4) For each cusp $\kappa \in \mathcal{P}$ the growth condition $f(g_\kappa a(y) g) = O(N(y)^a)$ as $y \to \infty$ is satisfied for some a (depending on f and κ).

If $d > 1$ this condition is automatically satisfied (see 2.2.2 below).

We shall often say *automorphic form* for "χ-automorphic form".

The automorphic forms in §1 are obtained by taking $d = 1$, $I = \mathbb{Z} = O_\mathbb{Q}$, $q = 0$. Then we have functions on $\mathrm{SL}_2(\mathbb{R})/\mathrm{SO}_2(\mathbb{R}) \cong \mathfrak{H}$. The terms in the Fourier expansions (1.1) and (1.4) are eigenfunctions of the Casimir operator C, and have the polynomial growth required in condition (A4).

Classical holomorphic Hilbert modular forms h on $\mathfrak{H}^d$ of weight $q \in (2\mathbb{Z})^d$ transform according to $h(\gamma z) = \chi(d_\gamma) \left(\prod_j (c_j z_j + d_j)^{q_j} \right) h(z)$. Taking

$$(2.9) \quad f(n(x) a(y) k) = \prod_j y_j^{q_j/2} h(x + iy) \phi_q(k)$$

gives an automorphic form of weight q in the above sense, with spectral parameter ν given by $\nu_j = \frac{q_j - 1}{2}$, $1 \leq j \leq d$.

Any function of weight q is determined by its values on $NA \cong \mathfrak{H}^d$. The operators C_j correspond to elliptic differential operators on $\mathfrak{H}$. This implies that each locally integrable function satisfying (A2) and (A3) in the sense of distributions is real analytic on G.

More generally, a χ-automorphic form is a finite sum of automorphic forms with weight and spectral parameter depending on the summand. So a χ-automorphic form is a real analytic function on G that satisfies (A1) and (A4) and is K-finite and $\mathcal{Z}(\mathfrak{g})$-finite. Here $\mathcal{Z}(\mathfrak{g})$ denotes the center of the enveloping algebra of the Lie algebra of G. It corresponds to the algebra of invariant differential operators generated by the C_j.

2.2.1. *Fourier expansion.* For any $r \in \mathbb{R}^d$

$$(2.10) \quad \chi_r(n(x)) := e^{2\pi i S(rx)}$$

is a character of N.

Any continuous function f on G satisfying (A1) has a Fourier expansion at ∞

$$(2.11) \quad f(ng) = \sum_{r \in O'} \chi_r(n) F_r f(g) \quad (n \in N),$$

$$F_r f(g) := \frac{1}{\text{vol}(\mathbb{R}^d \bmod O)} \int_{\mathbb{R}^d \bmod O} e^{-2\pi i S(rx)} f(n(x)g)\, dx$$
$$= \frac{1}{\text{vol}(\Gamma_N \backslash N)} \int_{\Gamma_N \backslash N} \chi_r(n)^{-1} f(ng)\, dn.$$

The volume $\text{vol}\bigl(\mathbb{R}^d \bmod O\bigr)$ is equal to $\sqrt{|D_F|}$, where D_F is the discriminant of F over $\mathbb{Q}$. The *complementary ideal*
$$O' := \{x \in F \;:\; S(x\xi) \in \mathbb{Z} \text{ for all } \xi \in O\}$$
is a fractional ideal in F; it is equal to the inverse of the different of O over $\mathbb{Z}$.

If the function f also satisfies (A3), the Fourier terms $F_r f$ are also eigenfunctions of the Casimir operators. Together with the growth condition (A4), this implies that for $r \neq 0$ the Fourier term $F_r f$ is a multiple of the following function:

$$(2.12) \qquad W_q(r,\nu; na(y)k) := \chi_r(n)\phi_q(k) \prod_{j=1}^d W_{q_j \text{sign}(r_j)/2, \nu_j}(4\pi|r_j|y_j).$$

2.2.2. Götzky-Koecher principle. Proposition 4.9 in [14] states that in the classical context all holomorphic automorphic forms have polynomial growth if the degree d of F over $\mathbb{Q}$ is larger than one.

The proof of Proposition 2.2.2 in [7] shows that this principle holds more generally; it generalizes easily to the present situation.

2.2.3. Eisenstein series. A well known method of constructing functions satisfying the automorphic transformation rule is to start with a function h on G that satisfies $h(\gamma g) = \chi(\gamma)h(g)$ for γ in a subgroup $\tilde{\Gamma}$ of Γ, and to consider the series $\sum_{\gamma \in \tilde{\Gamma}\backslash \Gamma} \chi(\gamma)^{-1} h(\gamma g)$. If this series converges absolutely, it provides a function satisfying (A1).

The Eisenstein series is an instance where this works. We take $\tilde{\Gamma} = \Gamma_\kappa$ with $\kappa \in \mathcal{P}_\chi$. The function h is given by

$$(2.13) \qquad h_q^\kappa(\nu,\mu; g_\kappa n(x)a(y)k) = \prod_j y_j^{1/2+\nu+i\mu_j} \phi_q(k),$$

with $\nu \in \mathbb{C}$, and μ lying in a shifted lattice $\Lambda_{\kappa,\chi}$ in the hyperspace $S = 0$ in $\mathbb{R}^d$ determined by the condition that $h_q^\kappa(\nu,\mu;\gamma g) = \chi(\gamma) h_q^\kappa(\nu,\mu;g)$ for all $\gamma \in \Gamma_\kappa$. The *Eisenstein series at the cusp* κ

$$(2.14) \qquad E_q(\kappa,\chi;\nu,i\mu;g) := \sum_{\gamma \in \Gamma_\kappa \backslash \Gamma} \chi(\gamma)^{-1} h_q^\kappa(\nu,\mu;\gamma g)$$

converges absolutely for $\text{Re}\,\nu > \frac{1}{2}$ and defines an automorphic form of weight q with spectral parameter $(\nu + i\mu_1, \ldots, \nu + i\mu_d)$. As a function of ν it has a meromorphic continuation to $\mathbb{C}$ that, in the present context, is known to have no singularities with $\text{Re}\,\nu \geq 0$, except for a first order pole at $\nu = \frac{1}{2}$ if $\chi = 1$, $\mu = 0$, $q = 0$.

2.3. Spectral decomposition.

2.3.1. Weight spaces. Let $L^2(\Gamma \backslash G, \chi)$ be the Hilbert space of classes of measurable functions on G that satisfy (A1) almost everywhere on G, and for which

$$(2.15) \qquad \int_{\Gamma \backslash G} |f(g)|^2\, dg < \infty.$$

This integral gives the square of the norm $\|f\|$ in $L^2(\Gamma \backslash G, \chi)$. The scalar product is

$$(2.16) \qquad \langle f_1, f_2 \rangle := \int_{\Gamma \backslash G} f_1(g) \overline{f_2(g)}\, dg.$$

We shall use this notation also when the integral converges absolutely, but f_1 or f_2 is not square integrable.

Right translation R_g, given by $\left(R_g f\right)(x) = f(xg)$, leaves $L^2(\Gamma \backslash G, \chi)$ invariant. For a weight q satisfying (2.7), let $L^2(\Gamma \backslash G, \chi)_q$ be the closed subspace of $L^2(\Gamma \backslash G, \chi)$ in which K acts by the character ϕ_q. So the elements of $L^2(\Gamma \backslash G, \chi)_q$ satisfy (A1) and (A2) almost everywhere on G.

Elements of $L^2(\Gamma \backslash G, \chi)_q$ are determined by their values on $NA \cong \mathfrak{H}^d$. In this way, we may identify $L^2(\Gamma \backslash G, \chi)_q$ with the space of classes of functions $h : \mathfrak{H}^d \to \mathbb{C}$ satisfying

$$\int_{\Gamma \backslash \mathfrak{H}^d} |h(z)|^2 \, d\mu(z) < \infty, \tag{2.17}$$

$$h(\gamma \cdot z) = \chi(d) h(z) \prod_{j=1}^{d} e^{iq_j \arg(c_j z_j + d_j)} \qquad \text{for all } \gamma = \begin{pmatrix} a & b \\ c & d \end{pmatrix} \in \Gamma. \tag{2.18}$$

There is a pitfall in the normalization: If h corresponds to

$$f : n(x) \, a(y) \, k \mapsto h(x + iy) \phi_q(k),$$

then the integral in (2.17) is equal to $2\|f\|^2$. Indeed, if $\mathcal{F}_{\mathfrak{H}} \subset \mathfrak{H}^d$ is a fundamental domain for Γ in $\mathfrak{H}^d$, then the following set is a fundamental domain for Γ in G:

$$\mathcal{F}_G := \{ n(x) \, a(y) \, k \, : \, x + iy \in \mathcal{F}_{\mathfrak{H}}, \, k \in \mathcal{F}_K \}, \tag{2.19}$$

where $\mathcal{F}_K \subset K$ is a fundamental domain for $\{1, m(-1)\} \backslash K$.

The spectral theory of automorphic forms gives a decomposition of $L^2(\Gamma \backslash G, \chi)_q$ as an orthogonal direct sum

$$L_q^{2,\text{discr}}(\Gamma \backslash G, \chi) \oplus L_q^{2,\text{cont}}(\Gamma \backslash G, \chi).$$

The space $L^{2,\text{discr}}(\Gamma \backslash G, \chi)_q$ has an orthonormal basis $\{f_j\}$, where each f_j is a square integrable automorphic form of weight q, with spectral parameter $\nu^{(j)} \in \mathbb{C}^d$. For each $\nu \in \mathbb{C}^d$ there are only finitely many values of j with $\nu^{(j)} = \nu$. If $q = 0$ and $\chi = 1$, one of the f_j's is a constant function. In the present context, we know that all other f_j's are *cusp forms*: $F_0 f_j = 0$. For noncongruence subgroups of $SL_2(\mathbb{R})$ (case $d = 1$), there may be nonconstant noncuspidal square integrable automorphic forms, given by residues of Eisenstein series. These automorphic forms will contribute to the sum formula.

The subspace $L^{2,\text{cont}}(\Gamma \backslash G, \chi)_q$ can be described by integrals of Eisenstein series. For the sum formula, it suffices to know that for bounded functions f_1, f_2 in $L^2(\Gamma \backslash G, \chi)_q$, the projections f_1^{cont} and f_2^{cont} onto the space $L^{2,\text{cont}}(\Gamma \backslash G, \chi)_q$ satisfy

$$\begin{aligned}\left\langle f_1^{\text{cont}}, f_2^{\text{cont}} \right\rangle \\ = \sum_{\kappa \in \mathcal{P}_\chi} c_\kappa \sum_{\mu \in \mathcal{L}_\kappa} \int_{-\infty}^{\infty} \left\langle f_1, E_q(\kappa, \chi; iy, i\mu) \right\rangle \overline{\left\langle f_2, E_q(\kappa, \chi; iy, i\mu) \right\rangle} \, dy,\end{aligned} \tag{2.20}$$

for suitable constants c_κ. These constants depend on the geometry of the fundamental domain near the cusps, hence c_κ does not depend on the character χ.

These results are well known but it is hard to pinpoint a reference where they are stated in the present context. They follow from [31], or [17], but there the theory is given for much more general groups. The present situation of an algebraic group with $\mathbb{Q}$-rank one is at the bottom of the induction procedure in [31]. The proof of the spectral decomposition of $L^2(\Gamma \backslash G, \chi)_q$ follows the same lines as for $L^2(\Gamma_0 \backslash \mathfrak{H})$, where Γ_0 is a cofinite volume discrete

	spectr. parameter	weights
trivial representation	0	0
even unitary principal series	$\nu \in i[0, \infty)$	$2\mathbb{Z}$
odd unitary principal series	$\nu \in i(0, \infty)$	$2\mathbb{Z} + 1$
complementary series	$\nu \in \left(0, \tfrac{1}{2}\right)$	$2\mathbb{Z}$
holomorphic discrete series	$\nu = \tfrac{b-1}{2},\ b \in \mathbb{N},\ b \geq 2$	$b + 2\mathbb{N}_0$
antiholomorphic discr. series	$\nu = \tfrac{b-1}{2},\ b \in \mathbb{N},\ b \geq 2$	$-b - 2\mathbb{N}_0$
holo. mock discr. series	$\nu = 0$	$1 + 2\mathbb{N}_0$
antiholo. mock discr. series	$\nu = 0$	$-1 - 2\mathbb{N}_0$

TABLE 1. Irreducible unitary representations of $\mathrm{SL}_2(\mathbb{R})$.

subgroup of $\mathrm{PSL}_2(\mathbb{R})$ acting on the upper half plane $\mathfrak{H}$. See, e.g., Chap. 4–7 in [**21**], [**2**], and the survey [**48**], Chap. 4–5.

2.3.2. *Automorphic representations.* The product structure $G = \mathrm{SL}_2(\mathbb{R})^d$ induces a product structure for complexified Lie algebra: $\mathfrak{g} = (\mathfrak{sl}_2)^d$. In the j-th factor there are elements $\mathbf{E}_j^+$ and $\mathbf{E}_j^-$ that correspond to the following differential operators on G:

$$(2.21) \qquad \mathbf{E}_j^\pm = e^{\pm 2i\vartheta_j}\left(\pm 2iy_j\partial_{x_j} + 2y_j\partial_{y_j} \mp i\partial_{\vartheta_j}\right),$$

with Iwasawa coordinates (x, y, ϑ).

If f is a differentiable function on G of weight q (see (2.6)), then $\mathbf{E}_j^\pm f$ has weight $q \pm 2\varepsilon_j$, where $\varepsilon_j \in \mathbb{R}^d$ has coordinate 1 at position j and zeros elsewhere. These operators preserve the transformation behavior (A1). As they commute with the Casimir operators C_j, the spectral parameters are preserved as well. In particular, these operators act on automorphic forms.

By repeated application of the $\mathbf{E}_j^\pm$ to a given automorphic form f of weight q with spectral parameter ν, we arrive at a basis of the $\mathfrak{g}$-module $\mathcal{U}(\mathfrak{g})f$, where $\mathcal{U}(\mathfrak{g})$ is the universal enveloping algebra of $\mathfrak{g}$. If, moreover, f is square integrable, then $\mathcal{U}(\mathfrak{g})f \subset L^2(\Gamma\backslash G, \chi)$ is an irreducible $(\mathfrak{g}, K)$-module, since f has a given fixed weight and a fixed spectral parameter. The corresponding representation ϖ of $\mathfrak{g}$ is the tensor product $\bigotimes_j \varpi_j$ of d irreducible unitary representations of $\mathrm{SL}_2(\mathbb{R})$. The components ν_j of the spectral parameter of f, and the q_j that occur in ϖ_j, are determined by the unitary dual of $\mathrm{SL}_2(\mathbb{R})$, see Table 1. In principle, there is the freedom $\nu_j \mapsto -\nu_j$ in the spectral parameter. In the table we have fixed our choice. If $\chi = 1$, the constant functions form a representation where every ϖ_j is the trivial representation. For all other ϖ occurring in $L^2(\Gamma\backslash G, \chi)$, none of the factors ϖ_j can be the trivial representation. This extends Proposition 4.11 in [**14**].

To see this, we consider an automorphic form f in the space of ϖ, where the factor ϖ_1 is the trivial representation. So f depends only on $(g_2, \ldots, g_d)$. From the fact that $O \subset \mathbb{R}^d$ projects onto a dense subset of $\mathbb{R}^{d-1}$, it follows from $f(n(x+\zeta)a(y)k) = f(n(x)a(y)k)$ for all $\zeta \in O$, that $f(nak)$ does not depend on n. So $f = F_0 f$, and

$$f(a(y)k(\vartheta)) = e^{iS(q\vartheta)} \prod_{2 \leq j \leq d,\, \nu_j \neq 0} \left(a_j y_j^{1/2+\nu_j} + b_j y_j^{1/2-\nu_j}\right)$$
$$\cdot \prod_{1 \leq j \leq d,\, \nu_j = 0} y_j^{1/2}\left(a_j + b_j \log y_j\right).$$

One checks that such a function can be Γ-invariant only if it is constant.

2. PRELIMINARIES

Let f be a square integrable automorphic form of weight q with spectral parameter ν. The space $\mathcal{U}(\mathfrak{g})f$ is contained in $f \subset L^{2,\text{discr}}(\Gamma\backslash G, \chi)$. It is irreducible for the action of $\mathfrak{g}$. It is not invariant under right translation by general elements of G. However, the closure in $L^{2,\text{discr}}(\Gamma\backslash G, \chi)$ is an irreducible unitary G-module, also called an *automorphic representation*. We shall denote the representation of $\mathfrak{g}$ on $\mathcal{U}(\mathfrak{g})f$ and the corresponding representation of G on $\overline{\mathcal{U}(\mathfrak{g})f}$ by the same symbol.

2.3.3. *Central character.* We have seen that in an irreducible representation ϖ of G the weights are in one class of $\mathbb{Z}^d$ mod $2\mathbb{Z}^d$. Let us represent this class by $\xi \in \{0,1\}^d$. The center M acts on the irreducible representation ϖ by multiplication by the *central character*

$$(2.22) \qquad m(\zeta_1, \ldots, \zeta_d) \mapsto \prod_{j=1}^{d} \zeta_j^{\xi_j} \qquad (\zeta_1, \ldots, \zeta_d) \in \{1, -1\}^d.$$

A necessary condition for ϖ to occur in $L^2(\Gamma\backslash G, \chi)$ is

$$(2.23) \qquad \chi(m(-1)) = (-1)^{S(\xi)} = (-1)^{\xi_1 + \cdots + \xi_d}.$$

We have a decomposition into orthogonal G-invariant subspaces

$$(2.24) \qquad L^2(\Gamma\backslash G, \chi) = \bigoplus_{\xi} L^2_{\xi}(\Gamma\backslash G, \chi),$$

where ξ runs through the elements of $\{0,1\}^d$ satisfying (2.23), and M acts on the space $L^2_{\xi}(\Gamma\backslash G, \chi)$ by multiplication by the character specified by ξ.

We shall develop a sum formula for each of the summands $L^2_{\xi}(\Gamma\backslash G, \chi)$.

2.3.4. *Fourier coefficients.* The operators $\mathbf{E}^{\pm}_j$ commute with the Fourier term operators F_r in (2.11). The functions W_q in (2.12) satisfy

$$(2.25) \qquad \mathbf{E}^{\pm}_j W_q(r, \nu) = W_{q \pm 2\varepsilon_j}(r, \nu) \cdot \begin{cases} -2 & \text{if } \pm r_j > 0, \\ \left(\frac{1}{2}(q_j \pm 1)^2 - 2\nu_j^2\right) & \text{if } \pm r_j < 0. \end{cases}$$

These equalities follow from differentiation relations for Whittaker functions, see formulas (2.4.21) and (2.4.24) in [**44**].

We turn to a consequence of (2.25) for the Fourier expansion of automorphic forms. Suppose that the j-th component ϖ_j of the square integrable automorphic representation ϖ is a discrete series representation, or a mock discrete series representation. So $\nu_j = \frac{b-1}{2}$ with $b \in \mathbb{N}$, $b \equiv q_j$ mod 2. Then we can find an automorphic form $f \neq 0$ for ϖ, for which the weight q satisfies $q_j = \pm b$ (b in the holomorphic case, and $-b$ in the antiholomorphic case). At place j, the weight $\pm b$ is the lowest, respectively highest, weight occurring in the representation. So $\mathbf{E}^{\mp}_j f = 0$. Let $r \in \mathcal{O}'$, $r \neq 0$. We have $F_r f = a_r W_q(r, \nu_{\varpi})$ for some $a_r \in \mathbb{C}$. Now $0 = F_r \mathbf{E}^{\mp}_j f = a_r \mathbf{E}^{\mp}_j W_q(r, \nu_{\varpi})$. From (2.25) it follows that $a_r = 0$ if $\pm r_j < 0$. This generalizes the classical result that holomorphic Hilbert cusp forms have Fourier expansions with terms of totally positive order; see Proposition 4.9 in [**14**]. If a form is holomorphic only at some real places, then we have only Fourier terms with an order that is positive at those places.

If the j-th factor is in the holomorphic discrete series, then $W_{b_j, \nu_j}(t) = t^{b_j/2} e^{-\frac{t}{2}}$. So indeed, the j-th factor of W_b corresponds to $y_j^{b_j/2}$ times a holomorphic function on $\mathfrak{H}$.

The discrete series case just considered is the sole occurrence of $\mathbf{E}^{\pm}_j f = 0$ for square integrable automorphic forms. From the structure of square integrable automorphic representations $\varpi = \bigotimes_j \varpi_j$ and relation (2.25) it follows that the Fourier terms of non-zero order for various weights determine each other. The Fourier coefficients do not depend on the weight in an essential way. In fact, we can arrange an orthogonal basis $(\psi_{\varpi,q})$ of the space of each ϖ that satisfies the following conditions:

i) For each weight q occurring in ϖ, the function $\psi_{\varpi,q}$ is an automorphic form of weight q with spectral parameter ν_ϖ. Its norm satisfies

$$(2.26) \qquad \|\psi_{\varpi,q}\|^2 = n(q, \nu_\varpi) := \prod_{j=1}^{d} n(q_j, \nu_{\varpi,j}),$$

$$n(q,\mu) := 1 \qquad \text{if } \operatorname{Re}\mu = 0,\ q \in \mathbb{Z},$$

$$:= \frac{\Gamma\left(\frac{1}{2} - \mu + \frac{q}{2}\right)}{\Gamma\left(\frac{1}{2} + \mu + \frac{q}{2}\right)} = \frac{\Gamma\left(\frac{1}{2} - \mu - \frac{q}{2}\right)}{\Gamma\left(\frac{1}{2} + \mu - \frac{q}{2}\right)} \qquad \text{if } 0 < \mu < \frac{1}{2},\ q \in 2\mathbb{Z},$$

$$:= \left(\frac{q-b}{2}\right)! \Big/ \left(\frac{b+q-2}{2}\right)! \qquad \text{if } 2\mu+1 = b \in \mathbb{N},$$
$$q \geq b,\ q \equiv b \bmod 2,$$

$$:= \left(\frac{|q|-b}{2}\right)! \Big/ \left(\frac{b+|q|-2}{2}\right)! \qquad \text{if } 2\mu+1 = b \in \mathbb{N},$$
$$q \leq -b,\ q \equiv b \bmod 2.$$

ii) $\mathbf{E}_j^\pm \psi_{\varpi,q} = \left(1 + 2\nu_{\varpi,j} \pm q_j\right) \psi_{\varpi,q\pm 2\varepsilon_j}$ for $j = 1, \ldots, d$.

It may seem unnatural that not all $\psi_{\varpi,q}$ have norm 1. However, the present choice avoids square roots in formulas later on. The relation in ii) is satisfied by the Eisenstein series. We impose it for the $\psi_{\varpi,q}$ as well. This fixes the $\psi_{\varpi,q}$ up to a scalar factor not depending on q. We derive from $\langle \mathbf{E}_j^\pm \psi_{\varpi,q}, \psi_{\varpi,q\pm 2\varepsilon_j}\rangle + \langle \psi_{\varpi,q}, \mathbf{E}_j^\mp \psi_{\varpi,q\pm 2\varepsilon_j}\rangle = 0$, the condition $\frac{n(q\pm 2\varepsilon_j, \nu)}{n(q,\nu)} = \frac{1-2\bar\nu_j \pm q_j}{1+2\nu_j \pm q_j}$. (Here ε_j denotes the j-th unit vector in $\mathbb{R}^d$.) This fixes the norms up to a constant factor. With the choice of the norms as in i), the choice of the $\psi_{\varpi,q}$ has still the freedom of a common factor of absolute value one.

With this normalization in place, we define the Fourier coefficients $c^r(\varpi)$ of ϖ with non-zero order $r \in O$ by

$$(2.27) \qquad F_r \psi_{\varpi,q} = c^r(\varpi) d^r(q, \nu_\varpi) W_q(r, \nu_\varpi) \qquad \text{for all weights in } \varpi,$$

$$(2.28) \qquad d^r(q,\nu) := \frac{1}{\sqrt{2^d |D_F N(r)|}} \prod_{j=1}^{d} \frac{e^{\pi i q_j}}{\Gamma\left(\frac{1}{2} + \nu_j + \frac{q_j}{2}\operatorname{sign}(r_j)\right)}.$$

A computation based on (2.25) and the relation in ii) show that $c^r(\varpi)$ does not depend on q. The gamma factor does not cause zeros for any q_j that really occurs in ϖ_j. We have already seen that if ϖ_j is of discrete (or mock discrete) series type with $\nu_{\varpi,j} = \frac{b-1}{2}$, then

$$(2.29) \qquad c^r(\varpi) \neq 0 \implies r_j q_j > 0.$$

If we multiply all $\psi_{\varpi,q}$ by $u \in \mathbb{C}$, $|u| = 1$, then the $c^r(\varpi)$ are multiplied by the same factor.

In the case of a representation from the unitary principal series or the complementary series, we have for each $l = 1, \ldots, d$ the possibility of replacing ν_l by $-\nu_l$. The corresponding change of the basis elements can be arranged by $\tilde\psi_{\varpi,q} = \Gamma\left(\frac{1}{2} + \nu_l - \frac{q_l}{2}\right)\Gamma\left(\frac{1}{2} - \nu_l - \frac{q_l}{2}\right)^{-1} \psi_{\varpi,q}$. This leads to the following Fourier coefficients

$$(2.30) \qquad \tilde c^r(\varpi) = (-1)^{\xi_l} c^r(\varpi).$$

The same considerations lead to the following description of the Fourier coefficients of the Eisenstein series: If $r \in O'$, $r \neq 0$, $q \in \xi + 2\mathbb{Z}^d$, $\xi \in \{0, 1\}^d$:

$$F_r E_q(\kappa, \chi; \nu, i\mu) = D_\xi^r(\kappa, \chi; \nu, i\mu) d^r(q, \nu + i\mu) W_q(r, \nu + i\mu) \tag{2.31}$$

Here $\nu \in \mathbb{C}$ is identified with $(\nu, \nu, \ldots, \nu) \in \mathbb{C}^d$. In the same way as for the $c^r(\varpi)$, one checks that the $D_\xi^r(\kappa, \chi)$ do not depend on the weight.

2.4. Kloosterman sums.
The sum formula Theorem 3.21 will relate Fourier coefficients to Kloosterman sums associated to the number field F.

We define for $c \in O \setminus \{0\}$, $r, r' \in O' \setminus \{0\}$, and a character χ of $(O \bmod (c))^*$:

$$S_\chi(r, r'; c) := \sum_{a \bmod (c)}^* \chi(a) e^{2\pi i S((r'a + r\tilde{a})/c)}, \tag{2.32}$$

where a runs over representatives of $(O \bmod (c))^*$, and $\tilde{a}a \equiv 1 \bmod (c)$. This definition generalizes the Kloosterman sums over $\mathbb{Q}$ in (1.7). Some easy properties are

$$|S_\chi(r, r'; c)| \leq |N(c)|, \tag{2.33}$$

$$S_\chi(r, r'; -c) = \chi(-1) S_\chi(r, r'; c), \tag{2.34}$$

$$S_\chi(r', r; c) = S_{\chi^{-1}}(r, r'; c), \tag{2.35}$$

$$\overline{S_\chi(r, r'; c)} = \chi(-1) S_{\chi^{-1}}(r, r'; c) = \chi(-1) S_\chi(r', r; c). \tag{2.36}$$

To derive the sum formula, we do not need any estimate of Kloosterman sums. The absolute convergence of the sum in (3.4) will follow from the absolute convergence of the Poincaré series. However, it will turn out in §3.5.1 that a nontrivial estimate of Kloosterman sums allows us to enlarge the class of test functions.

The classical Kloosterman sums, see (1.7), satisfy the *Weil-Salié estimate*

$$S(m, n; c) \ll_{m,n,\delta} c^{\frac{1}{2} + \delta} \tag{2.37}$$

for each $\delta > 0$. See Salié, [42], Weil, [52], and also Estermann, [13].

We have worked out the generalization to Kloosterman sums $S_1(r', r; c)$ over an arbitrary algebraic number field, see §5 of [5], following Estermann's approach. Later, we found that Gundlach had indicated a similar extension of Estermann's work in §4 of [16].

The discussion in [5] goes through for any number field F. For non-zero c in the ring of integers of F and for r, r' in the complement, the Kloosterman sums are

$$S(r, r'; c) = \sum_{a \bmod (c)}^* e^{2\pi i \text{Tr}_{F/\mathbb{Q}}((r'a + r\tilde{a})/c)}, \tag{2.38}$$

like in (2.32), with $\chi = 1$. Theorem 10 in [5] gives

$$|S(r, r'; c)| \leq C_F \sqrt{N_{r,r'}(c)}\, 2^{\text{pr}(c)} |N(c)|^{1/2}, \tag{2.39}$$

with:
- An explicit constant C_F depending on the number field F.
- pr(c) the number of prime ideals dividing (c). In the last remark of §5.2 in [5] one finds that $2^{\text{pr}(c)} = O_\delta\left(|N(c)|^\delta\right)$ for each $\delta > 0$.
- Let v_P denote the valuation associated to the prime ideal P.

$$N_{r,r'}(c) = \prod_{P, v_P(c) > 0} NP^{\min(v_P(r), v_P(r'), v_P(c) - d_P)}. \tag{2.40}$$

The d_P describe the different: $\mathfrak{D}_{F/\mathbb{Q}} = \prod_P P^{d_P}$.

For the present situation, with F totally real, we shall derive a similar estimate for the Kloosterman sums with a character χ modulo I. For $c \neq 0$, $c \in I$, and $r, r' \in O'$, let us consider the characters φ and ψ of $O/(c)$ given by

$$(2.41) \quad \varphi(a) = e^{2\pi i S(r'a/c)}, \quad \psi(a) = e^{2\pi i S(ra/c)}.$$

So (2.32) takes the form $S_\chi(r, r'; c) = \sum_{a \in (O/(c))^*} \chi(a)\varphi(a)\psi(\tilde{a})$, with $a\tilde{a} \equiv 1 \mod (c)$. The influence of the choice of the generator c of the ideal (c) now goes into the definition of φ and ψ. This is a special case of the generalized Kloosterman sum

$$(2.42) \quad S_\chi(\varphi, \psi; J) := \sum_{a \in (O/J)^*} \chi(a)\varphi(a)\psi(\tilde{a}),$$

with $\tilde{a}a \equiv 1 \mod J$, defined for each ideal $J \subset I$ and characters φ and ψ of the additive group O/J.

We apply the method used in [5] to these generalized Kloosterman sums. We expand J as a product $\prod_{i=1}^q P_i^{m_i}$ of prime ideals P_i in O. The ring O/J is isomorphic to the direct product of rings $\prod_{i=1}^q O/P_i^{m_i}$. This corresponds with product expansions $\varphi = \bigotimes_i \varphi_i$, $\psi = \bigotimes_i \psi_i$, and $\chi_i = \bigotimes_i \chi_i$, where φ_i and ψ_i are characters of the additive group $O/P_i^{m_i}$, and χ_i are characters of $\left(O/P_i^{m_i}\right)^*$. The assumption that χ is a character modulo I implies that $\chi_i = 1$ if P_i does not divide I. In the corresponding product formula

$$(2.43) \quad S_\chi(\varphi, \psi; J) = \prod_{i=1}^q S_{\chi_i}\left(\varphi_i, \psi_i; P_i^{m_i}\right),$$

we estimate the factors such that $P_i \mid I$ trivially, see (2.33). For the other factors, Proposition 9 in [5] gives:

$$(2.44) \quad \left|S_1\left(\varphi_i, \psi_i; P_i^{m_i}\right)\right| \leq c(P_i) N(P_i)^{m_i - N_i/2},$$

with N_i minimal such that $P_i^{N_i}$ is contained in both $\ker \varphi_i$ and $\ker \psi_i$. The constants $c(P_i)$ are equal to 2 unless $P_i \mid (2)$.

Let us now take $J = (c)$, and φ and ψ as in (2.41). In §5.2 of [5], we have shown that the N_i are given by

$$(2.45) \quad N_i = \max\left(0, -v_{P_i}(r') - d_{P_i} + v_{P_i}(c), -v_{P_i}(r) - d_{P_i} + v_{P_i}(c)\right).$$

This gives

$$(2.46) \quad \left|S_\chi(r', r; c)\right|$$
$$\leq \prod_{i, P_i \mid I} N(P_i)^{m_i} \cdot \prod_{i, P_i \nmid I} c(P_i) N(P_i)^{\frac{1}{2} v_{P_i}(c) + \frac{1}{2} \max(v_{P_i}(r'), v_{P_i}(r))}$$
$$\ll_{F, I, r, r'} 2^{\mathrm{pr}(c)} \prod_{P_i, P_i \mid I} N(P_i)^{m_i} \prod_{P_i, P_i \nmid I} N(P_i)^{m_i/2} |N(c/I_c)|^{\frac{1}{2}}.$$

We take into account that $2^{\mathrm{pr}(c)} = O_\delta\left(|N(c)|^\delta\right)$, and obtain the following *Weil type estimate*:

PROPOSITION 2.1. *Let $r, r' \in O' \setminus \{0\}$, and let χ be a character of $(O/I)^*$. For $c \in I \setminus \{0\}$, we write $(c) = I_c J_c$, with $I_c = \prod_{P, v_P(I) > 0} P^{v_P(c)}$, where P runs over prime ideals of O, and J_c is relatively prime to I.*

For each $\delta > 0$:

$$(2.47) \quad S_\chi(r', r; c) = O_{F, r', r, \delta}\left(N(I_c) N(J_c)^{\frac{1}{2} + \delta}\right).$$

This result will be used in §3.5.4 to enlarge the class of test functions to include holomorphic functions on a narrow strip.

2.5. Poincaré series. We apply the method sketched at the start of §2.2.3 with $\tilde{\Gamma}$ equal to Γ_N. The character χ of Γ is as in §2.1.1, with $\chi\begin{pmatrix} a & b \\ c & d \end{pmatrix} = \chi(d)$.

Under suitable conditions on the function h on $\Gamma_N\backslash G$, the *Poincaré series*

$$(2.48) \qquad Ph(g) := \sum_{\gamma \in \Gamma_N\backslash \Gamma} \chi(\gamma)^{-1} h(\gamma g)$$

converges absolutely. Clearly, then Ph satisfies (A1). The main difficulty is that for degree $d > 1$, the convergence requires handling a sum over the infinitely many units. In this respect, we recall Lemma 4.4 in [9], which gives:

LEMMA 2.2. *Let $\alpha, \beta \in \mathbb{R}$, with $\alpha + \beta > 0$. If $f : (\mathbb{R}^*)^d \to \mathbb{C}$ satisfies*

$$(2.49) \qquad |f(y)| \le \prod_{j=1}^{d} \min\left(p_j |y_j|^\alpha, q_j |y_j|^{-\beta}\right),$$

with $p_j > 0$, $q_j > 0$, then

$$(2.50) \qquad \sum_{\zeta \in O^*} f(\zeta y) \ll \left(1 + \left|\log |N(y)| + \frac{1}{\alpha + \beta} \log \frac{P}{Q}\right|\right)^{d-1}$$
$$\cdot \min\left(P|N(y)|^\alpha, Q|N(y)|^{-\beta}\right),$$

with $P = \prod_j p_j$, $Q = \prod_j q_j$.

The quotient $\Gamma_N \backslash \Gamma_\infty$ is represented by the matrices $\begin{pmatrix} \zeta & 0 \\ 0 & \zeta^{-1} \end{pmatrix} = a(\zeta^2) m(\zeta)$ with $\zeta \in O^*$. Thus we obtain:

LEMMA 2.3. *Let $h : \Gamma_N \backslash G \to \mathbb{C}$ satisfy*

$$(2.51) \qquad |h(na(y)k)| \le C \prod_{j=1}^{d} \min\left(y_j^\alpha, y_j^{-\beta}\right),$$

with $\alpha, \beta \in \mathbb{R}$, $\alpha + \beta > 0$, and $C \ge 0$. Then there exists $C_{\alpha,\beta} \ge 0$ such that

$$\sum_{\gamma \in \Gamma_N\backslash\Gamma_\infty} |h(\gamma n a(y) k)|$$
$$\le C C_{\alpha,\beta} \left(1 + |\log N(y)|^{d-1}\right) \min\left(N(y)^\alpha, N(y)^{-\beta}\right).$$

This lemma is relevant only if $d > 1$.

LEMMA 2.4. *Assume that the function $h : \Gamma_N \backslash G \to \mathbb{C}$ is continuous and satisfies (2.51) with $\alpha > 1$ and $\alpha + \beta > 0$. Then the Poincaré series in (2.48) converges absolutely. For each suitably small $\varepsilon > 0$:*

$$(2.52) \qquad Ph(na(y)k) \ll_{\alpha,\beta,\varepsilon} \max\left(N(y)^{1-\alpha+\varepsilon}, N(y)^{-\beta+\varepsilon}\right) \quad \text{as } N(y) \to \infty,$$

and at other cusps $\kappa \in \mathcal{P} \setminus \{\infty\}$:

$$(2.53) \qquad Ph(g_\kappa na(y)k) \ll_{\alpha,\varepsilon} N(y)^{1-\alpha+\varepsilon} \quad \text{as } N(y) \to \infty.$$

If $\beta > 0$, respectively $\beta > -\frac{1}{4}$, then Ph is bounded, respectively Ph represents an element of $L^2(\Gamma\backslash G, \chi)$.

PROOF. Let $f(g) = \sum_{\gamma \in \Gamma_N \backslash \Gamma_\infty} \chi(\gamma)^{-1} h(\gamma g)$. The previous lemma shows that this defines a continuous function on G satisfying $f(\gamma g) = \chi(\gamma) f(g)$ for all $\gamma \in \Gamma_\infty$, and

$$(2.54) \qquad f(na(y)k) \ll_{\alpha,\beta,\varepsilon} \begin{cases} N(y)^{\alpha-\varepsilon} & \text{as } N(y) \downarrow 0, \\ N(y)^{-\beta+\varepsilon} & \text{as } N(y) \to \infty. \end{cases}$$

Let us take $\varepsilon > 0$ such that $\alpha - \varepsilon > 1$. The convergence of $Ph(g) = \sum_{\gamma \in \Gamma_\infty \backslash \Gamma} \chi(\gamma)^{-1} f(\gamma g)$ follows from the convergence of the Eisenstein series $E_0(\infty, 1; \sigma, 0; g)$, with $\sigma = \alpha - \varepsilon - \frac{1}{2}$. Moreover, if we omit the term with $\gamma \in \Gamma_\infty$ from this Eisenstein series, we are left with a sum that is $O\left(N(y)^{1/2-\sigma}\right)$ as $N(y) \to \infty$. Thus we obtain

$$(2.55) \qquad Ph(na(y)k) = f(na(y)k) + O\left(N(y)^{1-\alpha+\varepsilon}\right) \quad \text{as } N(y) \to \infty.$$

This gives (2.52). There is only the term $O\left(N(y)^{1-\alpha+\varepsilon}\right)$ at cusps that are not equivalent to ∞. This gives (2.53).

The function Ph is bounded near the cusps inequivalent to ∞, hence square integrable. Square integrability near ∞ is determined by the growth of the term $f(na(y)k)$. So $\beta > -\frac{1}{4}$ is a sufficient condition. For boundedness, one can argue in a similar way. □

3. Derivation of the spectral sum formula

We are ready to start the derivation of the first main result in this paper, Theorem 3.21.

As stated before, most proofs of the sum formula are based on the computation of the scalar product of two square integrable Poincaré series. That is the approach we shall follow.

3.1. Inner product of Poincaré series.
Let $r \in O'$, $r \neq 0$, and let $q \in \mathbb{Z}^d$ be a weight satisfying (2.7). We shall use Poincaré series that are built with a continuous function $h = h_q^r$ satisfying the transformation rule $h_q^r(nak) = \chi_r(n) h_q^r(a) \phi_q(k)$ and the estimate (2.51) with $\alpha > 1$, $\beta > 0$. The sum formula is based on the computation of the scalar product of two such Poincaré series Ph_q^r and $Ph_q^{r'}$ in two different ways.

The first way is based on the spectral decomposition. We use the relation between Poincaré series and Fourier terms. For any continuous $f \in L^2(\Gamma \backslash G, \chi)_q$:

$$(3.1) \qquad \langle Ph_q^r, f \rangle = \int_{\Gamma_N \backslash G} h_q^r(g) \overline{f(g)} \, dg$$
$$= \sqrt{|D_F|} \int_A h_q^r(a) \overline{F_r f(a)} \, |a|^{-1} \, da.$$

In the special case that f is an automorphic form of weight q with spectral parameter ν, we find with (2.12) that $\langle Ph_q^r, f \rangle$ is a multiple of

$$\int_A h_q^r(a) \overline{W_q(r, \nu; a)} \, |a|^{-1} \, da.$$

Here we see the appearance of a Whittaker transform, which we shall study in §3.2. In the case $d = 1$, $q = 0$, this is the Kontorovitch-Lebedev transformation; see (5.14.14) in [**32**].

In the spectral description of $\langle Ph_q^r, Ph_q^{r'} \rangle$, we shall arrive at expressions involving Fourier coefficients of automorphic forms and Whittaker transforms of h_q^r and $h_q^{r'}$.

In the second way to compute the scalar product of Poincaré series, we shall apply (3.1) with $f = Ph_q^{r'}$. Note that the parameters r and r' determining the character of N may differ, but not the weight q; otherwise the scalar product in (3.1) would vanish. Using the absolute convergence, we find

(3.2) $\quad F_r\left(Ph_q^{r'}\right)(a(y))$
$$= \frac{1}{\sqrt{|D_F|}} \sum_{\gamma \in \Gamma_N \backslash \Gamma} \chi(\gamma)^{-1} \int_{\Gamma_N \backslash N} \chi_r(n(x))^{-1} h_q^{r'}(\gamma n(x) a(y)) \, dx.$$

This expression has to be integrated against h_q^r to obtain another way to compute the scalar product $\langle Ph_q^r, Ph_q^{r'}\rangle$. In this subsection, we expand the expression in (3.2).

Let us first consider the subsum over $\Gamma_N \backslash \Gamma_\infty$, see (2.8).

$$\frac{1}{\sqrt{|D_F|}} \sum_{\zeta \in O^*} \chi(\zeta) \int_{\Gamma_N \backslash N} \chi_r(n)^{-1} h_q^{r'}\left(a(\zeta^2) m(\zeta) na(y)\right) dn$$

$$= \frac{1}{\sqrt{|D_F|}} \sum_{\zeta \in O^*} \chi(\zeta) \int_{\mathbb{R}^d \bmod O} e^{-2\pi i S(rx)+2\pi i S(r'\zeta^2 x)}$$
$$\cdot h_q^{r'}\left(a(y\zeta^2) m(\zeta)\right) dx$$

(3.3) $\quad = \sum_{\zeta \in O^*, r=\zeta^2 r'} \chi(\zeta) h_q^{r'}\left(a(y\zeta^2) m(\zeta)\right).$

Note that $\Gamma_\infty = \Gamma \cap NAM$. So here we have considered the elements of Γ that lie in the small cell in the *Bruhat decomposition* $G = NAM \sqcup NAMwN$, $w := \begin{pmatrix} 0 & 1 \\ -1 & 0 \end{pmatrix}$.

For $\gamma = \begin{pmatrix} a & b \\ c & d \end{pmatrix}$ in the big cell $NAMwN$, we write $\gamma = n\left(\frac{a}{c}\right) a(c^{-2}) m(-c) wn\left(\frac{d}{c}\right)$. The sum over $\gamma \in \Gamma_N \backslash (\Gamma \cap NAMwN)$ amounts to letting c run over the non-zero elements of I, a over representatives of $(O \bmod (c))^*$, and d over the $d \in O$ such that $ad \in 1 + (c)$. We write $d = \tilde{a} + \zeta c$, where $\tilde{a}$ is one solution of $\tilde{a}a \equiv 1 \bmod (c)$, and ζ runs over O. As χ is a character of $(O/I)^*$, this gives, with the notation $\sum'_{c \in I}$ for the sum over $c \in I \setminus \{0\}$:

$$\frac{1}{\sqrt{|D_F|}} \sum_{c \in I}' \sum_{a \bmod (c)}^* \sum_{\zeta \in O} \chi(a) \int_{x \in \mathbb{R}^d \bmod O} e^{-2\pi i S(rx)}$$
$$\cdot h_q^{r'}\left(n\left(\tfrac{a}{c}\right) a(c^{-2}) m(-c) wn\left(\tfrac{\tilde{a}}{c} + \zeta + x\right) a(y)\right) dx$$

$$= \frac{1}{\sqrt{|D_F|}} \sum_{c \in I}' \sum_{a \bmod (c)}^* \chi(a) \int_{\mathbb{R}^d} e^{-2\pi i S(r(x-\tilde{a}/c))+2\pi i S(r'a/c)}$$
$$\cdot h_q^{r'}\left(wa(c^{-2}) m(-c) n(x) a(y)\right) dx$$

$$= \frac{1}{\sqrt{|D_F|}} \sum_{c \in I}' S_\chi(r, r'; c)$$

(3.4) $\quad \cdot \int_N \chi_r(n)^{-1} h_q^{r'}\left(w^{-1} a(c^2) m(c) na(y)\right) dn.$

As χ is a character of $(O \bmod I)^*$, it induces a character of $(O \bmod (c))^*$, for which the Kloosterman sum $S_\chi(r, r'; c)$ makes sense.

Let us check that these series and integrals converge absolutely. If we form the Poincaré series $P_1|h_q^r(g)|$, with χ in (2.48) replaced by 1, the estimates in Lemma 2.4 hold. The scalar product $\langle P_1|h_q^r|, P_1|h_q^{r'}|\rangle$ is finite. Equation (3.1) and the computations above show that it is equal to

(3.5) $\quad \int_A |h_q^r(a)| \left(\sum_{\varepsilon \in O^*} \left| h^{r'}(aa(\varepsilon^2) m(\varepsilon)) \right| \right.$

$$+ \sum_{c \in I}{}' \#(O/(c))^* \int_N \left|h_q^r\left(w^{-1}a(c^2)m(c)na\right)\right| dn \Big) |a|^{-1} da.$$

Thus, we have obtained a majorant for the expressions in (3.3) and (3.4) themselves, and for their integrals on A against h_q^r.

The absolute convergence of these Poincaré series is based on Lemma 2.2 and on the convergence of Eisenstein series. We do not need any estimate of Kloosterman sums to prove convergence.

3.2. Auxiliary test functions. In order to obtain a sum formula like that in §1, where the test function depends on the spectral parameter, it is convenient not to use h_q^r and h_q^r as the auxiliary functions, but their spectral transforms. That brings us to the spectral decomposition of the space $L^2(N\backslash G, r)$ of functions on G transforming on the left according to the character $\chi_r : n(x) \mapsto e^{2\pi i S(rx)}$ of N. We shall obtain this as a consequence of the Whittaker transform applied to the function $a \mapsto h_q^r(a)$ on A, as in [4]. See [49] for a treatment for general reductive groups.

3.2.1. *Whittaker transform.* The space $N\backslash G$ is a product of d copies of the space $N\backslash SL_2(\mathbb{R})$, with $N = \{n(x) : x \in \mathbb{R}\}$. The Whittaker transform has a corresponding product structure. Many computations can be carried out locally.

We use the same notations for objects at one place, as for the corresponding objects obtained as a product over all real places.

The spectral data are functions on sets of the following form:

DEFINITION 3.1. Let $0 < \tau < 1$.

If $\xi \in \{0, 1\}$ represents a class of $\mathbb{Z}$ mod $2\mathbb{Z}$, put

(3.6) $$D(0, \tau) := \{v \in \mathbb{C} : |\operatorname{Re} v| \leq \tau\} \cup \left(\tfrac{1}{2} + \mathbb{N}_0\right),$$
$$D(1, \tau) := \{v \in \mathbb{C} : |\operatorname{Re} v| \leq \tau\} \cup \mathbb{N}.$$

If $\xi \in \{0, 1\}^d$, put

(3.7) $$D(\xi, \tau) := \prod_{j=1}^d D(\xi_j, \tau).$$

Next we define the spectral data themselves.

DEFINITION 3.2. Let $\tfrac{1}{2} < \tau < 1$.

For $u \in \mathbb{Z}$, take $\xi \in \{0, 1\}$ such that $u \equiv \xi \bmod 2$. By $H_{u,\tau}$ we denote the space of functions η on $D(\xi, \tau)$ that are holomorphic and even on $|\operatorname{Re} v| \leq \tau$, and satisfy

(3.8) $$\eta(v) \ll e^{-\frac{\pi}{2}|\operatorname{Im} v|}(1 + |\operatorname{Im} v|)^{-a} \quad \text{for each } a \in \mathbb{R},$$

and

(3.9) $$\eta\left(\tfrac{b-1}{2}\right) = 0 \quad \text{if } b \equiv u \bmod 2, \, b > u.$$

Let $q \in \mathbb{Z}^d$ satisfy $q \equiv \xi \bmod 2\mathbb{Z}^d$ and (2.7). Let $r \in O'$, $r \neq 0$. Put $u_j = q_j \operatorname{sign}(r_j)$, for $j = 1 \cdots, d$. The space $H_{q,\tau}^r$ consists of the functions η on $D(\xi, \tau)$ of the form

$$\eta(v) = \prod_{j=1}^d \eta_j(v_j),$$

with $\eta_j \in H_{u_j,\tau}$.

Note that in the local situation, $\eta(\frac{b-1}{2})$ is non-zero at only finitely many elements of $\frac{\xi-1}{2} + \mathbb{Z}$. So (3.8) is a statement concerning the growth on the strip $|\operatorname{Re}\nu| \leq \tau$. If $u \leq 2$, there are no points $\frac{b-1}{2} > \tau$, and η is effectively a function on the strip.

We have used Greek letters for the elements of $H_{u,\tau}$. We shall mostly adhere to the following policy as long as it is practical: Greek letters denote spectral data (or functions depending on spectral data), whereas Latin letters denote functions on the group G or on spaces related to G.

Basic is the Whittaker transform in the next local result. It connects elements of $H_{u,\tau}$ to functions on $(0, \infty)$.

THEOREM 3.3. *Let $\frac{1}{2} < \tau < 1$, $u \in \mathbb{Z}$. For $\eta \in H_{u,\tau}$, define for $y \in (0, \infty)$:*

$$(3.10) \quad w_u\eta(y) := \frac{1}{4\pi i} \int_{\operatorname{Re}\nu=0} \eta(\nu) W_{u/2,\nu}(y) \left| \frac{\Gamma\left(\frac{1}{2} + \nu - \frac{u}{2}\right)}{\Gamma(2\nu)} \right|^2 d\nu$$

$$+ \sum_{b \equiv u \bmod 2,\, 1 < b \leq u} \eta\left(\frac{b-1}{2}\right) W_{u/2,(b-1)/2}(y) \frac{b-1}{\left(\frac{u-b}{2}\right)! \left(\frac{u+b-2}{2}\right)!}.$$

This defines a continuous function on $(0, \infty)$ satisfying

$$(3.11) \quad w_u\eta(y) \ll \min\left(y^{\frac{1}{2}+\tau}, y^{\frac{1}{2}-\tau}\right).$$

For any continuous function f on $(0, \infty)$ satisfying the bound in (3.11), the integral in

$$(3.12) \quad \omega_u f(\nu) := \int_0^\infty f(y) W_{u/2,\nu}(y) \frac{dy}{y^2}$$

converges absolutely for all ν satisfying $|\operatorname{Re}\nu| < \tau$, or $\nu = \frac{b-1}{2}$, with $b \equiv u \bmod 2$, $1 < b \leq u$. For these values of ν we have

$$(3.13) \quad \omega_u w_u \eta(\nu) = \eta(\nu)$$

for each $\eta \in H_{u,\tau}$.

See [4], Propositions 13.3.10 and 13.3.13, and their proofs. The idea is the same as in §5. Proposition 13.3.5 shows that ω_u extends unitarily to an isomorphism of Hilbert spaces

$$(3.14) \quad \omega_u : L^2\left((0, \infty), \frac{dy}{y^2}\right) \longrightarrow$$

$$L^2\left(i[0, \infty) \cup \left\{\frac{b-1}{2} : 1 < b \leq u,\, b \equiv u \bmod 2\right\}, d\lambda\right),$$

where the measure $d\lambda$ is given by $\left|\frac{\Gamma(\frac{1}{2}+\nu-\frac{u}{2})}{\Gamma(2\nu)}\right|^2 \frac{d\nu}{2\pi i}$ for $\nu \in i[0, \infty)$ and has point masses $\frac{b-1}{\left(\frac{u-b}{2}\right)!\left(\frac{u+b-2}{2}\right)!}$ at the points $\frac{b-1}{2}$.

At each place, we can view $(0, \infty)$ as $A = \{a(y) : y > 0\} \subset \operatorname{SL}_2(\mathbb{R})$, and associate a function on $\operatorname{SL}_2(\mathbb{R})$ to $w_u\eta$ for $\eta \in H_{u,\tau}$.

DEFINITION 3.4. *If $q \in \mathbb{Z}$, $r \in \mathbb{R} \setminus \{0\}$, and $\eta \in H_{q \operatorname{sign} r, \tau}$, put*

$$w_q^r \eta(n(x)a(y)k(\vartheta)) := e^{2\pi i r x} w_{q \operatorname{sign} r} \eta(4\pi |r|y) e^{iq\vartheta}.$$

If $q \in \mathbb{Z}^d$, $r \in \mathcal{O}'$, $r \neq 0$, and $\eta = \bigotimes_j \eta_j \in H_{q,\tau}^r$, define $w_q^r\eta$ on G by

$$w_q^r \eta(g) := \prod_{j=1}^d w_{q_j}^{r_j} \eta_j(g_j).$$

We shall use Poincaré series $Pw_q^r\eta$, with $\eta \in H_{q,\tau}^r$, in the proof of the sum formula. The convergence is guaranteed, as (3.11) implies that condition (2.51) is satisfied with $\alpha = \frac{1}{2} + \tau > 1$, and $\beta = \tau - \frac{1}{2} > 0$.

The factor $W_q(r, v)$ in the Fourier term in (2.27) contains a product of Whittaker functions. As indicated in (3.1), we shall integrate the Poincaré series $Pw_q^r\eta$ against automorphic forms. Then the following consequences of Theorem 3.3 will be useful:

Corollary 3.5. *For $q \in \mathbb{Z}^d$, $r \in O' \setminus \{0\}$, and $\eta \in H_{q,\tau}^r$:*

$$(3.15) \qquad \int_A w_q^r\eta(a) \overline{W_q(r, v; a)} |a|^{-1} \, da = (4\pi)^d |N(r)| \eta(v),$$

if for all v_j either $|\operatorname{Re} v_j| < \tau$, or $2v_j + 1 \equiv q_j \mod 2$, $1 < 2v_j + 1 \leq q_j \operatorname{sign}(r_j)$.

Corollary 3.6. *For $q \in \mathbb{Z}^d$, $r \in O' \setminus \{0\}$, and $\eta, \eta' \in H_{q,\tau}^r$:*

$$(3.16) \qquad \int_A w_q^r\eta(a) \overline{w_q^r\eta'(a)} |a|^{-1} \, da$$

$$= (4\pi)^d |N(r)| \prod_{j=1}^d \left(\int_0^{i\infty} \eta_j(v) \overline{\eta_j'(v)} \left| \frac{\Gamma\left(\frac{1}{2} + v - \frac{u_j}{2}\right)}{\Gamma(2v)} \right|^2 \frac{dv}{2\pi i} \right.$$

$$\left. + \sum_{1 < b \leq u_j} \eta_j\left(\frac{b-1}{2}\right) \overline{\eta_j'\left(\frac{b-1}{2}\right)} \frac{b-1}{\left(\frac{u_j-b}{2}\right)! \left(\frac{u_j+b-2}{2}\right)!} \right),$$

with $u_j = q_j \operatorname{sign} r_j$.

3.2.2. *Integral operator.* Let us turn to the integral in (3.4), with $h_q^r = w_q^r\eta$. It is the product of integrals of the following type:

$$(3.17) \qquad \int_{-\infty}^\infty e^{-2\pi i r_j x} w_{q_j \operatorname{sign} r_j'}^{r_j'} \eta_j\left(w^{-1} a(c^2) m(c) n(x) a(y)\right) dx,$$

depending on the j-th component of $r, r' \in O' \setminus \{0\}$ and $q \in \mathbb{Z}^d$.

In the investigation of the integral in (3.17), we work for the moment at one real place of F. We write $r, r' \in \mathbb{R}^*$ instead of r_j, r_j'; $\varepsilon = \operatorname{sign}(r)$, $\varepsilon' = \operatorname{sign}(r')$; $q \in \mathbb{Z}$ instead of q_j.

Let us consider for $\eta \in H_{q\varepsilon',\tau}$, the integral in (3.17):

$$(3.18) \qquad I(c, g) := \int_{-\infty}^\infty e^{-2\pi i r x} w_{q\varepsilon'}^{r'} \eta\left(w^{-1} a(c^2) m(c) n(x) a(y)\right) dx.$$

Our aim is to arrive at Theorem 3.8, which will give $I(c, g) = w_{q\varepsilon}^r \tilde{\eta}(g)$ with $\tilde{\eta} \in H_{q\varepsilon,\tau}$ depending on c. The method we follow is present in Chapter 13 of [**4**].

The spectral expansion of $w_{q\varepsilon'}^{r'}\eta$ is based on (3.10). Let us express the W-Whittaker function in (3.10) in terms of M-Whittaker functions, see p. 10 in [**44**].

$$W_{u,v}(y) = \sum_{\pm} \Gamma(\mp 2v) \Gamma\left(\tfrac{1}{2} \mp v - \tfrac{u}{2}\right)^{-1} M_{u, \pm v}(y)$$

as an identity of meromorphic functions in v,

$$W_{u, \frac{b-1}{2}}(y) = (-1)^{(u-b)/2} \left(\tfrac{u+b-2}{2}\right)! \, ((b-1)!)^{-1} M_{u, \frac{b-1}{2}}(y)$$

for $b \equiv u \mod 2$, $b \leq u$.

We insert these expressions into (3.10), use the symmetry of the integrand, and move the line of integration to $\operatorname{Re} v = \alpha \in \left(0, \tfrac{1}{2}\right)$:

$$(3.19) \quad w_u \eta(y) = \frac{1}{2\pi i} \int_{\operatorname{Re} v = \alpha} \eta(v) M_{u/2,v}(y) \frac{\Gamma\left(\tfrac{1}{2} + v - \tfrac{u}{2}\right)}{\Gamma(2v)} dv$$

$$+ \sum_{b \equiv u \bmod 2,\, 1 < b \leq u} \eta\left(\tfrac{b-1}{2}\right) M_{u/2,(b-1)/2}(y) \frac{(-1)^{(u-b)/2}}{(b-2)!\, \tfrac{u-b}{2}!}.$$

We reformulate this result in terms of $w_q^{r'} \eta$:

$$(3.20) \quad M_v^{r'}(n(x)a(y)k(\vartheta)) := e^{2\pi i r' x} M_{q\varepsilon'/2,v}(4\pi |r'| y) e^{iq\vartheta},$$

$$(3.21) \quad w_q^{r'} \eta(g) = \frac{1}{2\pi i} \int_{\operatorname{Re} v = \alpha} \eta(v) M_v^{r'}(g) \frac{\Gamma\left(\tfrac{1}{2} + v - \tfrac{q\varepsilon'}{2}\right)}{\Gamma(2v)} dv$$

$$+ \sum_{b \equiv q \bmod 2,\, 1 < b \leq \varepsilon' q} \eta\left(\frac{b-1}{2}\right) M_{(b-1)/2}^{r'}(g) \frac{(-1)^{(q\varepsilon'-b)/2}}{(b-2)!\, \tfrac{q\varepsilon'-b}{2}!}.$$

We want to insert this into the integral for $I(c,g)$ in (3.18), and interchange the order of integration.

An estimate based on

$$(3.22) \quad w^{-1} a(c^2) m(c) n(x) a(y)$$
$$= n\left(\tfrac{-x}{c^2(x^2+y^2)}\right) a\left(\tfrac{y}{c^2(x^2+y^2)}\right) k(\arg(c(x-iy)))$$

and the asymptotic behavior of the M-Whittaker function near 0 show that the integral

$$(3.23) \quad V_v(r,r',c;g) := \int_{\mathbb{R}} e^{-2\pi i r x} M_v^{r'}\left(w^{-1} a(c^2) m(c) n(x) g\right) dx$$

converges absolutely if $\operatorname{Re} v > 0$, and

$$(3.24) \quad V_v(r,r',c;a(y)) = O\left(y^{\tfrac{1}{2} - \operatorname{Re} v}\right) \qquad \text{as } y \to \infty.$$

We arrive at:

$$(3.25) \quad I(c,g) = \frac{1}{2\pi i} \int_{\operatorname{Re} v = \alpha} \eta(v) V_v(r,r',c;g) \frac{\Gamma\left(\tfrac{1}{2} + v - \tfrac{q\varepsilon'}{2}\right)}{\Gamma(2v)} dv$$

$$+ \sum_{b \equiv q \bmod 2,\, 1 < b \leq q\varepsilon'} \eta\left(\frac{b-1}{2}\right) V_{(b-1)/2}(r,r',c;g) \frac{(-1)^{(q\varepsilon'-b)/2}}{(b-2)!\, \tfrac{q\varepsilon'-b}{2}!}.$$

Thus, we are led to the task of computing $V_v(r,r',c)$. The behavior of the factor $M_{\varepsilon' q/2,v}(t)$ as $t \downarrow 0$ allows differentiation under the integral sign in (3.23). So $V_n(r,r',c;g)$ is an eigenfunction of the Casimir operator with eigenvalue $\tfrac{1}{4} - v^2$. We see that

$$V_v(r,r',c;n(x)gk(\vartheta)) = e^{2\pi i r x} V_v(r,r',c;g) e^{iq\vartheta}.$$

The space of eigenfunctions of the Casimir operator satisfying this transformation rule and (3.24), is spanned by

$$(3.26) \quad W_v^r(n(x)a(y)k) := e^{2\pi i r x} W_{\varepsilon q/2,v}(4\pi |r| y) e^{iq\vartheta}.$$

PROPOSITION 3.7. *For* $\operatorname{Re} \nu > 0$, $c, r, r' \in \mathbb{R} \setminus \{0\}$, $g \in SL_2(\mathbb{R})$:

$$V_\nu(r, r', c; g) = \frac{2\pi \Gamma(2\nu + 1) \sqrt{|r'/r|}}{\Gamma\left(\frac{1}{2} + \nu + \frac{\varepsilon q}{2}\right)} \frac{(-i \operatorname{sign} c)^q}{|c|} J_{2\nu}^{\varepsilon \varepsilon'}\left(\frac{4\pi \sqrt{|rr'|}}{|c|}\right) W_\nu^r(g),$$

where $J_a^1 = J_a$, $J_a^{-1} = I_a$, are respectively the J- and I-Bessel functions

(3.27) $$J_a^{\pm 1}(t) = \sum_{n=1}^{\infty} \frac{(\mp 1)^n}{n! \Gamma(a + 1 + n)} \left(\frac{t}{2}\right)^{a+2n}.$$

Remark 1. This result is given in [4], Chapter 6. As this is the place where Bessel functions enter the sum formula, it seems worthwhile to give a proof that relates it to the action of the Casimir operator on functions on the big cell in the Bruhat decomposition. The basic idea is present in [39].

Remark 2. Miatello and Wallach, [34], generalize this result to general Lie groups with $\mathbb{R}$-rank one, using the theory of Whittaker vectors on principal series representations. The essential point is loc. cit. (2.10), with the specialization to $SL_2(\mathbb{R})$ on p. 461, line 3 from bottom. Let us write $\chi(n(x)) = e^{2\pi i r' x}$. The operator $\varpi^\chi(\xi, \nu)$ assigns the function $(4\pi|r'|)^{-\frac{1}{2} - \nu} M_{\varepsilon' q}^{r'}$ to the standard vector of weight q in the corresponding principal series representation. See (1.2), loc. cit., for the operator $J_{\xi,\nu}^{\chi_1}$; it assigns a multiple of W_ν^r to the standard vector of weight q.

Remark 3. Cogdell and Piatetski-Shapiro give another approach in [11]. There, the appearance of Bessel functions arises from the Kirillov model of irreducible representations (see 3.1 of [11]). See also [38] (Motohashi).

PROOF. We have seen already that

(3.28) $$V_\nu(r, r', c; g) = C_\nu(r, r', c) W_\nu^r(g).$$

We have to determine C_ν.

From $M_\nu^r(g) = M_\nu^{\varepsilon'}(a(|r'|) g)$ it follows that

$$V_\nu(r, r', c) = V_\nu\left(r, \varepsilon', c/|r'|^{1/2}\right).$$

Also replacing x by $x/|r|$ in (3.23) gives

(3.29) $$V_\nu(r, r', c; g) = |r|^{-1} V_\nu\left(\varepsilon, \varepsilon', c/\sqrt{|rr'|}; a(|r|) g\right).$$

Now we consider the behavior as $|c| \to \infty$. From $M_{\varepsilon' q/2, \nu}(t) = t^{\frac{1}{2} + \nu} + O\left(t^{\frac{3}{2} + \operatorname{Re} \nu}\right)$ as $t \downarrow 0$, we see, using (3.22), that for $\operatorname{Re} \nu$ sufficiently large:

$$V_\nu(\varepsilon, \varepsilon', c; a(y))$$
$$\sim \left(\frac{4\pi y}{c^2}\right)^{\frac{1}{2} + \nu} \int_{-\infty}^{\infty} e^{-2\pi i \varepsilon x}(x^2 + y^2)^{-\frac{1}{2} - \nu} e^{iq \arg(x - iy)} dx \, e^{iq \arg c}$$
$$= \frac{(2\pi)^{1+2\nu}(-i)^q}{\Gamma\left(\frac{1}{2} + \nu + \frac{\varepsilon q}{2}\right)} (\operatorname{sign} c)^q |c|^{-1-2\nu} W_{\varepsilon q/2, \nu}(4\pi y) \qquad \text{as } |c| \to \infty.$$

This gives for $\operatorname{Re} \nu$ large:

(3.30) $$C_\nu(\varepsilon, \varepsilon', c) \sim \frac{(2\pi)^{1+2\nu}}{\Gamma\left(\frac{1}{2} + \nu + \frac{\varepsilon q}{2}\right)} (-i \operatorname{sign} c)^q |c|^{-1-2\nu} \qquad \text{as } |c| \to \infty.$$

To get an expression for C_ν, we turn to more complicated integrals than that in (3.23); we integrate over the big cell $NAMwN$ in the Bruhat decomposition, where we temporarily use the notation $G = \mathrm{SL}_2(\mathbb{R})$, with $N = \{n(x) : x \in \mathbb{R}\}$, $A = \{a(t) : t > 0\}$, $M = \{I, -I\}$. The Haar measure dg restricts to the measure $\frac{1}{\pi} dx \frac{dc}{|c|^3} dx_1$ in the coordinates $g \leftrightarrow n(x) a(c^2) m(c) w n(x_1)$.

Let f be a function on the big cell that transforms according to
$$f(n(x)\, g n(x_1)) = e^{2\pi i(\varepsilon x + \varepsilon' x_1)} f(g).$$
So f is determined by the function $\varphi(c) = f\!\left(a(c^2) m(c) w\right)$ on $\mathbb{R}^*$, which we suppose to be in $C_c^\infty(\mathbb{R}^*)$. The following computation shows that the convolution $f * M_\nu^{\varepsilon'}$ is well defined:

$$
\begin{aligned}
(3.31) \qquad f * M_\nu^{\varepsilon'}(g) &:= \int_{NAMwN/N} f(g_1) M_\nu^{\varepsilon'}(g_1^{-1} g)\, dg \\
&= \frac{1}{\pi} \int_{c\in\mathbb{R}^*} \int_{\mathbb{R}} f\!\left(n(x) a(c^2) m(c) w\right) \\
&\qquad \cdot M_\nu^{\varepsilon'}\!\left(w^{-1} a(c^{-2}) m(c) n(-x) g\right) dx\, \frac{dc}{|c|^3} \\
&= \frac{1}{\pi} \int_{\mathbb{R}^*} \varphi\!\left(\tfrac{1}{c}\right) V_\nu(\varepsilon, \varepsilon', c; g) |c|\, dc \\
&= \frac{1}{\pi} \int_{\mathbb{R}^*} \varphi\!\left(\tfrac{1}{c}\right) C_\nu(\varepsilon, \varepsilon', c) |c|\, dc\, W_\nu^\varepsilon(g).
\end{aligned}
$$

Let u be the distribution $\varphi \mapsto \int_{\mathbb{R}^*} \varphi(\tfrac{1}{c}) C_\nu(r, r', c)\, dc$ on $\mathbb{R}^*$. The Casimir operator C satisfies $(Cf) * M_\nu^{\varepsilon'} = \left(\tfrac{1}{4} - \nu^2\right) f * M_\nu^{\varepsilon'}$. A computation of C in coordinates on the big cell shows that the distribution u vanishes on test functions of the form
$$c \mapsto -\tfrac{1}{4c^2}\varphi''\!\left(\tfrac{1}{c}\right) + \tfrac{1}{4c}\varphi'\!\left(\tfrac{1}{c}\right) - \left(\tfrac{4\pi^2 \varepsilon \varepsilon'}{c^2} + \tfrac{1}{4} - \nu^2\right)\varphi\!\left(\tfrac{1}{c}\right),$$
with $\varphi \in C_c^\infty(\mathbb{R}^*)$. So $c \mapsto |c| C_\nu(\varepsilon, \varepsilon', \cdot)$ satisfies, as a distribution on $\mathbb{R}^*$, a regular differential equation with analytic coefficients, hence it is given by an analytic function. We write this function as $t \mapsto g(1/t)$. Then g satisfies:
$$(3.32) \qquad -\tfrac{1}{4} g''(c) - \tfrac{1}{4c} g'(c) - \left(4\pi^2 \varepsilon \varepsilon' - \tfrac{\nu^2}{c^2}\right) g(c) = 0.$$
This implies that g is of the form
$$g(c) = j(4\pi |c|, \mathrm{sign}(c)),$$
where $t \mapsto j(t, \pm 1)$ are solutions of the Bessel differential equation of order 2ν if $\varepsilon\varepsilon' > 0$, and of the corresponding modified Bessel differential equation if $\varepsilon\varepsilon' < 0$. The asymptotic behavior (3.30) implies:
$$(3.33) \qquad C_\nu(\varepsilon, \varepsilon', c) = 2\pi \frac{\Gamma(2\nu+1)}{\Gamma\!\left(\tfrac{1}{2} + \nu + \tfrac{\varepsilon q}{2}\right)} (-i\,\mathrm{sign}\, c)^q |c|^{-1} J_{2\nu}^{\varepsilon\varepsilon'}\!\left(\tfrac{4\pi}{|c|}\right),$$
first for $\mathrm{Re}\,\nu$ large, and by holomorphic continuation for $\mathrm{Re}\,\nu > 0$. With (3.28) and (3.29), we obtain the proposition. $\square$

For the computation of $I(c, g)$ we return to (3.25). We insert the expression for V_ν in Proposition 3.7. Let
$$\varphi(\nu) = \frac{2\pi \Gamma(2\nu + 1)\sqrt{|r'/r|}}{\Gamma\!\left(\tfrac{1}{2} + \nu + \tfrac{\varepsilon q}{2}\right)} \frac{(-i\,\mathrm{sign}\,c)^q}{|c|} J_{2\nu}^{\varepsilon\varepsilon'}\!\left(\tfrac{4\pi \sqrt{|rr'|}}{|c|}\right).$$

The function $v \mapsto \eta(v)\varphi(v)$ is holomorphic on $|\operatorname{Re} v| \leq \tau$ and has polynomial decay. Since $W_{\kappa,v}(t) \ll_{t,\kappa} e^{-\frac{\pi}{2}|\operatorname{Im} v|} |\operatorname{Im} v|^a$, $a \in \mathbb{R}$ for $|\operatorname{Re} v| \leq \tau$, we have absolute convergence of the resulting integral, and can move the line of integration back to $\operatorname{Re} v = 0$.

Let us divide the integrand by the factor $\left| \frac{\Gamma(\frac{1}{2}+v-\frac{\varepsilon q}{2})}{\Gamma(2v)} \right|^2$ in (3.10). We obtain

$$\eta(v)\varphi(v) \frac{\Gamma(-2v)\Gamma\left(\frac{1}{2} + v - \frac{\varepsilon' q}{2}\right)}{\Gamma\left(\frac{1}{2} + v - \frac{\varepsilon q}{2}\right)\Gamma\left(\frac{1}{2} - v - \frac{\varepsilon q}{2}\right)}$$

$$= \eta(v) \frac{\Gamma\left(\frac{1}{2} + v + \frac{\varepsilon q}{2}\right)}{\Gamma\left(\frac{1}{2} + v + \frac{\varepsilon' q}{2}\right)} \cdot \frac{-2\pi \cos \pi \left(v + \frac{\varepsilon q}{2}\right)}{\sin 2\pi v} \left|\frac{r'}{r}\right|^{1/2}$$

$$\cdot \frac{(-i \operatorname{sign} c)^q}{|c|} J^{\varepsilon\varepsilon'}_{2v}\left(\frac{4\pi |rr'|^{1/2}}{|c|}\right)$$

in front of $W^r_v(g)$. Symmetrization gives the following integral:

$$\frac{1}{4\pi i} \int_{\operatorname{Re} v = 0} \tilde{\eta}(v) W^r_v(g) \left| \frac{\Gamma(\frac{1}{2} + v - \frac{\varepsilon q}{2})}{\Gamma(2v)} \right|^2 dv,$$

where

(3.34) $\qquad \tilde{\eta}(v) := \eta(v) \frac{1}{2|r|} \frac{\Gamma\left(\frac{1}{2} + v + \frac{\varepsilon q}{2}\right)}{\Gamma\left(\frac{1}{2} + v + \frac{\varepsilon' q}{2}\right)} \frac{4\pi |rr'|^{1/2}}{|c|} k^{\varepsilon,\varepsilon'}_\xi \left(v, \frac{4\pi |rr'|^{1/2}}{c}\right),$

(3.35) $\qquad k^{\varepsilon,\varepsilon'}_\xi(v, t) := (-i\varepsilon \operatorname{sign} t)^\xi \cos \pi \left(v + \frac{\xi}{2}\right) \frac{(-\varepsilon\varepsilon')^\xi J^{\varepsilon\varepsilon'}_{-2v}(|t|) - J^{\varepsilon\varepsilon'}_{2v}(|t|)}{\sin 2\pi v}.$

Here we employ $\xi \in \{0, 1\}$, $\xi \equiv q \mod 2$. We have used the relation

$$\frac{\Gamma\left(\frac{1}{2} - v + \frac{\varepsilon q}{2}\right)}{\Gamma\left(\frac{1}{2} - v + \frac{\varepsilon' q}{2}\right)} = (\varepsilon\varepsilon')^q \frac{\Gamma\left(\frac{1}{2} + v + \frac{\varepsilon q}{2}\right)}{\Gamma\left(\frac{1}{2} + v + \frac{\varepsilon' q}{2}\right)}.$$

Note that

(3.36) $\qquad \overline{k^{\varepsilon,\varepsilon'}_\xi(v,t)} = (-1)^\xi k^{\varepsilon,\varepsilon'}_\xi(\bar{v}, t), \qquad k^{\varepsilon,\varepsilon'}_\xi(-v, t) = (\varepsilon\varepsilon')^\xi k^{\varepsilon,\varepsilon'}_\xi(v, t),$

$\qquad\qquad k^{\varepsilon,\varepsilon'}_\xi(v, -t) = (-1)^\xi k^{\varepsilon,\varepsilon'}_\xi(v, t).$

On the strip $|\operatorname{Re} v| \leq \tau$, the even function $\tilde{\eta}$ satisfies the growth condition (3.8), outside a neighborhood of possible poles. The function $v \mapsto k^{\varepsilon,\varepsilon'}_\xi(v, t)$ is holomorphic: At points $v = \frac{m}{2}$, $m \equiv \xi \mod 2$, we have $(\mp 1)^\xi J^{\pm 1}_{-2v}(t) = (\mp 1)^\xi (\mp 1)^m J^{\pm 1}_m(t) = J^{\pm 1}_m(t)$.

The gamma factors may produce singularities in the strip. Let us impose the condition that η is zero at the relevant points. Then $\tilde{\eta}$ behaves as an element of $H_{\varepsilon q, \tau}$ in the strip $|\operatorname{Re} v| \leq \tau$.

At points $v = \frac{b-1}{2}$, $b \equiv q \mod 2$, we have:

(3.37) $\qquad k^{\varepsilon,\varepsilon}_\xi\left(\frac{b-1}{2}, t\right) = (-1)^{(b-\xi)/2} (-i\varepsilon \operatorname{sign} t)^\xi J^1_{b-1}(|t|),$

$\qquad\qquad k^{\varepsilon,-\varepsilon}_\xi\left(\frac{b-1}{2}, t\right) = 0.$

This shows that $\tilde{\eta}$ as defined in (3.34) vanishes at points $\frac{b-1}{2}$ with $b \equiv q \mod 2$, $b > \varepsilon q$. So $\eta \mapsto \tilde{\eta}$ is a map from $H_{\varepsilon' q, \tau}$ to $H_{\varepsilon q, \tau}$.

We have arranged the definitions so that the integral in the right hand side of (3.21) is equal to the integral in

$$(3.38) \quad w_q^r \tilde{\eta}(g) = \frac{1}{4\pi i} \int_{\operatorname{Re} \nu = 0} \tilde{\eta}(\nu) W_\nu^r(g) \left| \frac{\Gamma\left(\frac{1}{2} + \nu - \frac{\varepsilon q}{2}\right)}{\Gamma(2\nu)} \right|^2 d\nu$$

$$+ \sum_{b \equiv \varepsilon q \bmod 2,\, 1 < b \leq \varepsilon q} \tilde{\eta}\left(\frac{b-1}{2}\right) W_{(b-1)/2}^r(g) \frac{b-1}{\left(\frac{\varepsilon q - b}{2}\right)! \left(\frac{\varepsilon q + b - 2}{2}\right)!}.$$

Let us consider the sum over b. First suppose that $\varepsilon' = -\varepsilon$. If $\varepsilon' q \leq 1$, then the sum in (3.21) is empty, and we have seen that $\tilde{\eta}\left(\frac{b-1}{2}\right) = 0$ for all terms that may occur in (3.38). If $\varepsilon' q > 0$, the sum in (3.21) is zero, for the same reason.

If $\varepsilon' = \varepsilon$, the sums in (3.21) and (3.38) are both present, or both empty. By (3.37) we see that they coincide.

In this way, we have arrived at the following result:

THEOREM 3.8. *Let $q \in \mathbb{Z}$, $r, r', c \in \mathbb{R}^*$, and $\frac{1}{2} < \tau < 1$; put $\varepsilon = \operatorname{sign} r$, $\varepsilon' = \operatorname{sign} r'$. Suppose that $\eta \in H_{\varepsilon' q, \tau}$ is such that*

$$\nu \mapsto \eta(\nu) \Gamma\left(\frac{1}{2} + \nu + \frac{\varepsilon q}{2}\right) \Gamma\left(\frac{1}{2} + \nu + \frac{\varepsilon' q}{2}\right)^{-1}$$

is holomorphic on $|\operatorname{Re} \nu| \leq \tau$.

Then $\tilde{\eta}$ as defined in (3.34) is an element of $H_{\varepsilon q, \tau}$, and

$$\int_{-\infty}^{\infty} e^{-2\pi i r x} w_q^{r'} \eta\left(w^{-1} a(c^2) m(c) n(x) g\right) dx = w_q^r \tilde{\eta}(g),$$

in the notation of Definition 3.4.

3.3. Restricted version of the sum formula. All preparations have been completed at this point. We start the development of the sum formula, by computing a scalar product of Poincaré series in two different ways.

3.3.1. *Fourier term orders and central character.* The sum formula depends on the choice of two Fourier term orders (n and m in §1) and a central character (trivial in §1).

We fix $r, r' \in \mathcal{O}' \setminus \{0\}$. These are orders of Fourier terms of automorphic forms on $\Gamma_1(I)$. As before, we denote the signs by $\varepsilon, \varepsilon' \in \{1, -1\}^d$, $\varepsilon_j = \operatorname{sign}(r_j)$.

The parity of the coordinates of the weights is determined by the central character, specified by $\xi \in \{0, 1\}^d$. It has to satisfy $\chi(-1) = (-1)^{S(\xi)} = (-1)^{\xi_1 + \cdots + \xi_d}$.

3.3.2. *Weights and auxiliary test functions.* For the preliminary sum formula in this subsection, we make some more choices.

We choose a *weight* $q \in \mathbb{Z}^d$ compatible with the central character: $q \equiv \xi \bmod 2\mathbb{Z}^d$. In principle, we can derive a preliminary sum formula for any weight. We shall let $|q|$ tend to infinity later on. To avoid the additional condition on η in Theorem 3.8, we prescribe the sign and put a condition on the size: $\varepsilon_j q_j > 2$ for all j.

For fixed $\tau \in \left(\frac{1}{2}, 1\right)$ and $\eta, \eta' \in H_{q,\tau}^r$, see Definition 3.2, we form the Poincaré series $P w_q^r \eta$ and $P w_q^r \eta'$, see (2.48) and Definition 3.4, which lie in $L^2(\Gamma \backslash G, \chi)_q$, see (3.11) and Lemma 2.4.

3.3.3. *Scalar product of Poincaré series, spectral description.* We apply the spectral decomposition in §2.3.1.

Discrete spectrum. For any square integrable automorphic representation ϖ that contains vectors with weight q, we have by (2.27) and (3.15):

(3.39) $$\langle Pw_q^r \eta, \psi_{\varpi,q} \rangle = \sqrt{|D_F|}\, \overline{c^r(\varpi)}\, \overline{d^r(q, v_\varpi)}\, (4\pi)^d |N(r)|\, \eta(v_\varpi).$$

According to (2.26) and (2.28), the contribution of $L^{2,\mathrm{discr}}(\Gamma\backslash G, \chi)_q$ to the scalar product $\langle Pw_q^r \eta, Pw_q^{r'} \eta' \rangle$ is:

(3.40) $$(8\pi^2)^d \sqrt{|N(rr')|} \sum_\varpi \overline{c^r(\varpi)}\, c^{r'}(\varpi)$$

$$\cdot \eta(v_\varpi) \overline{\eta'(v_\varpi)} \prod_{j=1}^d \frac{1}{\Gamma\left(\frac{1}{2} - v_{\varpi,j} + \frac{\varepsilon_j q_j}{2}\right) \Gamma\left(\frac{1}{2} + v_{\varpi,j} + \frac{\varepsilon'_j q_j}{2}\right)}.$$

For factors of complementary series type, we use that

$$\frac{1}{n(q_j, v_{\varpi,j})} = \Gamma\left(\frac{1}{2} + v_{\varpi,j} + \frac{\varepsilon_j q_j}{2}\right) \Gamma\left(\frac{1}{2} - v_{\varpi,j} + \frac{\varepsilon_j q_j}{2}\right)^{-1}.$$

The sum is over the ϖ with central character ξ that are visible in weight q. In particular, the trivial representation does not occur. Moreover, if the j-th factor of ϖ is in the discrete (or mock discrete) series, with $v_{\varpi,j} = \frac{b-1}{2}$, $b \geq 1$, then ϖ is present only if $\varepsilon_j = \varepsilon'_j$, and $\varepsilon_j q_j \geq b$. However, either $c^r(\varpi)$, $c^{r'}(\varpi)$, or $\Gamma\left(\frac{1}{2} + v_{\varpi,j} + \frac{\varepsilon'_j q_j}{2}\right)^{-1}$ is zero if $\varepsilon_j q_j < b$ or $\varepsilon'_j q_j < b$. So we need not indicate the range of ϖ under the summation sign; see (2.29).

In checking (3.40), one has to consider the three cases in (2.26) separately at a given place j, and use that $|q_j| = \varepsilon_j q_j = \varepsilon'_j q_j$ in the discrete series case. The fact that there are no case distinctions in the formulation of (3.40) is reassuring, and shows that the normalization of the $\psi_{\varpi,q}$ in (2.26) is adequate.

The functions η and η' are even in v_j on the strip $|\operatorname{Re} v_j| \leq \tau$, but the gamma factors are not. Let us replace v_j by $-v_j$, with $v_j \in i(0, \infty) \cup \left(0, \frac{1}{2}\right)$ (for $\xi_j = 0$), or $v_j \in i[0, \infty)$ (for $\xi_j = 1$). Then the product of gamma factors is multiplied by $(\varepsilon_j \varepsilon'_j)^{\xi_j}$, in agreement with relation (2.30). The case $v_j = 0$, $\xi_j = 1$, $\varepsilon_j = -\varepsilon'_j$ seems to give a contradiction. However, at least one of $c^r(\varpi)$ and $c^{r'}(\varpi)$ is zero under these assumptions; see (2.29).

Continuous spectrum. The estimates in Lemma 2.4 show that the scalar products with Eisenstein series in (2.20) converge absolutely for the relevant values of the spectral parameter. The computations go in the same way as for the discrete spectrum. The contribution of $L^{2,\mathrm{cont}}(\Gamma\backslash G, \chi)_q$ to $\langle Pw_q^r \eta, Pw_q^{r'} \eta' \rangle$ is

(3.41) $$(8\pi^2) \sqrt{|N(rr')|} \sum_{\kappa \in \mathcal{P}_\chi} c_\kappa \sum_{\mu \in \Lambda_{\kappa,\chi}} \int_{-\infty}^\infty \overline{D^r_\xi(\kappa, \chi; iy, i\mu)}$$

$$\cdot D^{r'}_\xi(\kappa, \chi; iy, i\mu)\, \eta(iy + i\mu)\, \overline{\eta'(iy + i\mu)}$$

$$\cdot \prod_{j=1}^d \frac{1}{\Gamma\left(\frac{1}{2} - iy - i\mu_j + \frac{\varepsilon_j q_j}{2}\right) \Gamma\left(\frac{1}{2} + iy + i\mu_j + \frac{\varepsilon'_j q_j}{2}\right)}\, dy.$$

3. DERIVATION OF THE SPECTRAL SUM FORMULA

Measure. We can formulate the description of the inner product in (3.40) with a measure $d\sigma_{\chi,\xi}^{r,r'}$ on the set

$$(3.42) \qquad Y_\xi := \prod_{j=1}^d Y_{\xi_j}, \qquad Y_\delta := \begin{cases} i[0,\infty) \cup \left(0, \tfrac{1}{2}\right] \cup \left(\tfrac{1}{2} + \mathbb{N}\right) & \text{if } \delta = 0, \\ i[0,\infty) \cup \mathbb{N} & \text{if } \delta = 1, \end{cases}$$

given by

$$(3.43) \qquad \int_{Y_\xi} h(\nu)\, d\sigma_{\chi,\xi}^{r,r'}(\nu) := \sum_\varpi \overline{c^r(\varpi)}\, c^{r'}(\varpi) h(\nu_\varpi) + 2 \sum_{\kappa \in \mathcal{P}_\chi} c_\kappa$$
$$\cdot \sum_{\mu \in \Lambda_{\kappa,\chi}} \int_0^\infty \overline{D_\xi^r(\kappa,\chi;i y, i\mu)}\, D_\xi^{r'}(\kappa,\chi; i y, i\mu) h(iy + i\mu)\, dy$$

for compactly supported continuous functions h. This measure is nonnegative if $r = r'$. If h is integrable for $d\sigma_{\chi,\xi}^{r,r}$ and for $d\sigma_{\chi,\xi}^{r',r'}$, then it is integrable for $d\sigma_{\chi,\xi}^{r,r'}$, and

$$(3.44) \qquad \int_{Y_\xi} |h(\nu)|\, \left|d\sigma_{\chi,\xi}^{r,r'}(\nu)\right|$$
$$\leq \left(\int_{Y_\xi} |h(\nu)|\, d\sigma_{\chi,\xi}^{r,r}(\nu)\right)^{1/2} \left(\int_{Y_\xi} |h(\nu)|\, d\sigma_{\chi,\xi}^{r',r'}(\nu)\right)^{1/2}.$$

The computations that we have carried out show that the function

$$(3.45) \qquad \vartheta_q^{r,r'}(\nu) := \eta(\nu)\, \overline{\eta'(\bar\nu)} \prod_{j=1}^d \frac{1}{\Gamma\left(\tfrac{1}{2} - \nu_j + \tfrac{\varepsilon_j q_j}{2}\right) \Gamma\left(\tfrac{1}{2} + \nu_j + \tfrac{\varepsilon_j q_j}{2}\right)}$$

on Y_ξ is integrable for all measures $d\sigma_{\chi,\xi}^{r,r'}$, and

$$(3.46) \qquad \left\langle Pw_q^r \eta, Pw_q^{r'} \eta' \right\rangle = (8\pi^2)^d \sqrt{|N(rr')|} \int_{Y_\xi} \vartheta_q^{r,r'}(\nu)\, d\sigma_{\chi,\xi}^{r,r'}(\nu).$$

We have used that $\eta'(\nu_j) = \eta'(\pm \bar\nu_j) = \eta'(\bar\nu_j)$ for $\nu_j \in Y_\xi$.

Equation (3.45) defines the function $\vartheta_q^{r,r'}$ on $D(\xi, \tau)$.

3.3.4. *Scalar product of Poincaré series, geometric description.* Now we apply the decomposition in (3.3) and (3.4). At the end of §3.1 we have indicated that the resulting integrals and sums converge absolutely. By (3.1), the scalar product of $Pw_q^r \eta$ and $Pw_q^{r'} \eta'$ is obtained by integrating $\sqrt{|D_F|} w_q^r \eta$ over A against the complex conjugates of the expressions in (3.3) and (3.4), where $h_q^r = w_q^r \eta$ and similarly for $h_q^{r'}$.

Delta term. In view of the $\delta_{m,n}$ in (1.6), we call the term arising from (3.3) the *delta term*. It is equal to:

$$(3.47) \qquad \sqrt{|D_F|} \int_A w_q^r \eta(a) \sum_{\zeta \in O^*,\, r = \zeta^2 r'} \chi(\zeta)^{-1} \overline{w_q^{r'} \eta'(a a(\zeta^2) m(\zeta))}\, |a|^{-1}\, da.$$

From $r = \zeta^2 r'$, it follows that $\varepsilon = \varepsilon'$. Definition 3.4 implies the equalities

$$w_q^{r'} \eta'\left(a(y\zeta^2)m(\zeta)\right) = \prod_j w_{\varepsilon_j q_j} \eta'_j\left(4\pi |r'_j| \zeta_j^2 y_j\right) (\operatorname{sign} \zeta_j)^{q_j}$$
$$= w_q^{r'} \eta'(a(y))\, \phi_q(m(\zeta)).$$

There are two choices for ζ, say ζ and $-\zeta$. From condition (2.7) it follows that $\chi(-\zeta)^{-1} \phi(m(-\zeta)) = \chi(\zeta)^{-1}\phi(m(\zeta))$. Let us define

$$\alpha(\chi, \xi; r, r') := \begin{cases} 2\chi(\zeta)^{-1}\phi_q(m(\zeta)) & \text{if } r = \zeta^2 r' \text{ with } \zeta \in O^*, \\ 0 & \text{otherwise.} \end{cases} \tag{3.48}$$

With (3.16), we can obtain an expression for the delta term. We want to write it as an integral of the function $\vartheta_q^{r,r'}$ in (3.45) against a measure not depending on q.

We define the measure $d\mathrm{Pl}_\delta$, with $\delta \in \{0, 1\}$, on Y_δ by

$$\int_{Y_0} h(\nu)\, d\mathrm{Pl}_0(\nu) := \int_0^{i\infty} h(\nu)(-4\pi\nu)\tan \pi\nu\, \frac{d\nu}{2\pi i} \tag{3.49}$$
$$+ \sum_{b>1,\, b\equiv 0 \bmod 2} (b-1) h\left(\tfrac{b-1}{2}\right),$$

$$\int_{Y_1} h(\nu)\, d\mathrm{Pl}_1(\nu) := \int_0^{i\infty} h(\nu)(4\pi\nu)\cot \pi\nu\, \frac{d\nu}{2\pi i}$$
$$+ \sum_{b>1,\, b\equiv 1 \bmod 2} (b-1) h\left(\tfrac{b-1}{2}\right),$$

If h is an even function on $i\mathbb{R}$, then the integrals $\int_0^{i\infty} \cdots \frac{d\nu}{2\pi i}$ can be replaced by $\int_{\mathrm{Re}\,\nu=0} \cdots \frac{d\nu}{4\pi i}$. The notation $d\mathrm{Pl}$ refers to the Plancherel measure for $\mathrm{SL}_2(\mathbb{R})$, see, e.g., [30], p. 174. The nonnegative measure $d\mathrm{Pl}_\xi$ on Y_ξ is defined as the product measure $d\mathrm{Pl}_\xi := \bigotimes_{j=1}^d d\mathrm{Pl}_{\xi_j}$.

In view of (3.48), the delta term is equal to

$$\sqrt{|D_F|}\, \alpha(\chi, \xi; r, r') \int_A w_q^r \eta(a) \overline{w_q^r \eta'\,(aa(|r/r'|))}\, da,$$

which needs to be considered only under the assumption $r_j r'_j > 0$ for all j. Definition 3.4 shows that we can replace $w_q^r \eta'(aa(|r/r'|))$ by $w_q^r \eta'(a)$. With (3.16), we obtain for the integral an expression of the form $(4\pi)^d |N(r)| \prod_j I_j$, with explicit factors I_j. Using that η'_j is even on the strip $|\mathrm{Re}\,\nu| \leq \tau$, we check that

$$I_j = \int_{Y_{\xi_j}} \frac{\eta_j(\nu) \overline{\eta'(\bar\nu)}}{\Gamma\left(\tfrac{1}{2} - \nu + \tfrac{\varepsilon_j q_j}{2}\right)\Gamma\left(\tfrac{1}{2} + \nu_j + \tfrac{\varepsilon_j q_j}{2}\right)}\, d\mathrm{Pl}_{\xi_j}(\nu).$$

A comparison with (3.45) shows that the delta term is given by:

$$\sqrt{|D_F|}\, \alpha(\chi, \xi; r, r')(4\pi)^d |N(r)| \int_{Y_\xi} \vartheta_q^{r,r'}(\nu)\, d\mathrm{Pl}_\xi(\nu). \tag{3.50}$$

If $\nu_j = \tfrac{b-1}{2}$, $|b| > \varepsilon'_j q_j$ for some j, then $\vartheta_q^{r,r'}(\nu) = 0$; see Definition 3.2.

The computations carried out imply that $\vartheta_q^{r,r'}$ is integrable for $d\mathrm{Pl}_\xi$. Actually, any measurable φ satisfying $\varphi(\nu) \ll \prod_{j=1}^d (1 + |\nu|)^{-a}$ for some $a > 2$ is integrable for $d\mathrm{Pl}_\xi$.

Bessel transform. The remaining contribution of the geometric side of the formula to the scalar product $\langle Pw_q^r\eta, Pw_q^r\eta'\rangle$ comes from the intersection of Γ with the big cell in the Bruhat decomposition. It is given by (3.1), taking into account (3.2) and (3.4). The absolute convergence allows taking the integral over A inside the sum over $c \in I \setminus \{0\}$. This brings us to consider first the quantity

$$\int_A w_q^r \eta(a) \overline{\int_N \chi_r(n)^{-1} w_q^r \eta'\,(w^{-1}a(c^2)m(c)na)\, dn}\, |a|^{-1}\, da. \tag{3.51}$$

The integral over $N \cong \mathbb{R}^d$ is the product of d integrals of the type considered in §3.2.2, see in particular Theorem 3.8. The additional condition on the factors of η' does not impose a restriction on η', as we have $\varepsilon_j q_j \geq \varepsilon'_j q_j$. Theorem 3.8 shows that the integral over N in (3.51) is equal to $w_q^r \tilde{\eta}'_c(a)$, with $\tilde{\eta}'_c \in H_{q,\tau}^r$ given by

$$(\tilde{\eta}'_c)_j(v) = \eta'_j(v) \frac{1}{2|r_j|} \frac{\Gamma\left(\frac{1}{2} + v + \frac{\varepsilon_j q_j}{2}\right)}{\Gamma\left(\frac{1}{2} + v + \frac{\varepsilon'_j q_j}{2}\right)} \frac{4\pi \sqrt{|r_j r'_j|}}{|c_j|} k_{\xi_j}^{\varepsilon_j, \varepsilon'_j}\left(v, \frac{4\pi \sqrt{|r_j r'_j|}}{c_j}\right),$$

with $k_{\xi_j}^{\varepsilon_j, \varepsilon'_j}$ given by (3.35). We obtain from (3.16) that the quantity in (3.51) is equal to

$$\int_A w_q^r \eta(a) \overline{w_q^r \tilde{\eta}'_c(a)} |a|^{-1} da$$

$$= (4\pi)^d |N(r)| \prod_{j=1}^d \frac{2\pi |r_j|^{1/2}}{|c_j| \, |r_j|^{1/2}} \left(\int_0^{i\infty} \eta_j(v) \overline{\eta'_j(v)} \right.$$

$$\cdot \frac{\Gamma\left(\frac{1}{2} - v + \frac{\varepsilon_j q_j}{2}\right) \Gamma\left(\frac{1}{2} + v - \frac{\varepsilon_j q_j}{2}\right) \Gamma\left(\frac{1}{2} - v - \frac{\varepsilon_j q_j}{2}\right)}{\Gamma\left(\frac{1}{2} - v + \frac{\varepsilon'_j q_j}{2}\right) \Gamma(2v) \Gamma(-2v)}$$

$$\cdot (-\varepsilon_j \varepsilon'_j)^{\xi_j} k_{\xi_j}^{\varepsilon_j, \varepsilon'_j}\left(v, \frac{4\pi \sqrt{|r_j r'_j|}}{c_j}\right) \frac{dv}{2\pi i}$$

$$+ \delta_{\varepsilon_j, \varepsilon'_j} \sum_{1 < b \leq \varepsilon_j q_j} \eta_j\left(\frac{b-1}{2}\right) \overline{\eta'_j\left(\frac{b-1}{2}\right)} \frac{b-1}{\frac{\varepsilon_j q_j - b}{2}! \frac{\varepsilon_j q_j + b - 2}{2}!}$$

$$\left. \cdot (-1)^{\xi_j} k_{\xi_j}^{\varepsilon_j, \varepsilon'_j}\left(\frac{b-1}{2}, \frac{4\pi \sqrt{|r_j r'_j|}}{c_j}\right) \right)$$

$$= \frac{(8\pi^2)^d}{|N(c)|} |N(rr')|^{1/2} \int_{Y_\xi} \vartheta_\xi^{r,r'}(v)$$

$$\cdot \prod_{j=1}^d (-1)^{\xi_j} k_{\xi_j}^{\varepsilon_j, \varepsilon'_j}\left(v_j, \frac{4\pi \sqrt{|r_j r'_j|}}{c_j}\right) d\mathrm{Pl}_\xi(v)$$

(3.52) $$= \frac{(8\pi^2)^d}{|N(c)|} |N(rr')|^{1/2} \chi(-1) \left(B_\xi^{\varepsilon, \varepsilon'} \vartheta_q^{r,r'}\right)\left(\frac{4\pi |rr'|^{1/2}}{c}\right),$$

with the Bessel transform

(3.53) $$B_\xi^{\varepsilon, \varepsilon'} \varphi(t) := \int_{Y_\xi} \varphi(v) \prod_{j=1}^d \left(k_{\xi_j}^{\varepsilon_j, \varepsilon'_j}(v_j, t_j)\right) d\mathrm{Pl}_\xi(v)$$

$$= \prod_j B_{\xi_j}^{\varepsilon_j, \varepsilon'_j} \varphi_j(t_j),$$

(3.54) $$B_{\xi_j}^{\varepsilon_j, \varepsilon'_j} \varphi_j(t) := \int_{Y_{\xi_j}} \varphi_j(v) k_{\xi_j}^{\varepsilon_j, \varepsilon'_j}(v, t) \, d\mathrm{Pl}_{\xi_j}(v).$$

We have used that the local integrals have an even integrand on $i\mathbb{R}$. Our computations have shown that the integral in (3.53) converges absolutely for $\varphi = \vartheta_q^{r,r'}$.

Let us give the local Bessel transforms more explicitly. For $\eta, \eta' \in \{1, -1\}$:

(3.55) $$B_0^{\eta,\eta}\varphi(t) = -\frac{i}{2}\int_{\operatorname{Re}\nu=0} \varphi(\nu)\frac{J_{2\nu}(|t|) - J_{-2\nu}(|t|)}{\cos \pi \nu} \nu \, d\nu$$
$$+ \sum_{b\geq 2,\, b\equiv 0 \bmod 2} (-1)^{b/2}(b-1)\varphi\left(\tfrac{b-1}{2}\right) J_{b-1}(|t|),$$

(3.56) $$B_1^{\eta,\eta}\varphi(t) = -\tfrac{\eta}{2}\operatorname{sign}(t)\int_{\operatorname{Re}\nu=0} \varphi(\nu)\frac{J_{2\nu}(|t|) + J_{-2\nu}(|t|)}{\sin \pi \nu} \nu \, d\nu$$
$$- i\eta \operatorname{sign}(t) \sum_{b\geq 3,\, b\equiv 1 \bmod 2} (-1)^{(b-1)/2}(b-1)\varphi\left(\tfrac{b-1}{2}\right) J_{b-1}(|t|),$$

(3.57) $$B_0^{\eta,-\eta}\varphi(t) = \tfrac{2i}{\pi} \int_{\operatorname{Re}\nu=0} \varphi(\nu) K_{2\nu}(|t|) \nu \sin \pi\nu \, d\nu,$$

(3.58) $$B_1^{\eta,-\eta}\varphi(t) = \tfrac{2}{\pi}\eta \operatorname{sign}(t) \int_{\operatorname{Re}\nu=0} \varphi(\nu) K_{2\nu}(|t|) \nu \cos \pi\nu \, d\nu.$$

Kloosterman term. In the light of (3.1) and (3.4), this leads to the following absolutely convergent contribution coming from the big cell of the Bruhat decomposition:

$$\frac{1}{\sqrt{|D_F|}} \sideset{}{'}\sum_{c\in I} \overline{S_\chi(r,r';c)} \sqrt{|D_F|}\, \frac{(8\pi^2)^d}{|N(c)|}\, |N(rr')|^{1/2}$$
$$\cdot \chi(-1)\left(B_\xi^{\varepsilon,\varepsilon'} \vartheta_q^{r,r'}\right)\left(\tfrac{4\pi|rr'|^{1/2}}{c}\right)$$

(3.59) $$= (8\pi^2)^d |N(rr')|^{1/2} \sideset{}{'}\sum_{c\in I} \frac{S_\chi(r',r;c)}{|N(c)|} \left(B_\xi^{\varepsilon,\varepsilon'} \vartheta_q^{r,r'}\right)\left(\tfrac{4\pi|rr'|^{1/2}}{c}\right).$$

The change from $S_\chi(r,r';c)$ to $S_\chi(r',r;c)$ is due to complex conjugation, see (2.36).

For functions f on $(\mathbb{R}^*)^d$, we define the sum of Kloosterman sums

(3.60) $$K_\chi^{r',r}(f) := \sideset{}{'}\sum_{c\in I} \frac{S_\chi(r',r;c)}{|N(c)|} f\left(\tfrac{4\pi|rr'|^{1/2}}{c}\right).$$

It converges absolutely for $f = B_\xi^{\operatorname{sign}(r),\operatorname{sign}(r')} \vartheta_q^{r,r'}$.

3.3.5. *Restricted sum formula.* We have obtained:

PROPOSITION 3.9. *Let $\tfrac{1}{2} < \tau < 1$. For all functions ϑ on Y_ξ of the form*

$$\vartheta(\nu) = \eta(\nu)\overline{\eta'(\bar\nu)} \prod_{j=1}^{d} \frac{1}{\Gamma\left(\tfrac{1}{2} - \nu_j + \tfrac{\varepsilon_j q_j}{2}\right)\Gamma\left(\tfrac{1}{2} + \nu_j + \tfrac{\varepsilon'_j q_j}{2}\right)},$$

with $q \equiv \xi \bmod 2\mathbb{Z}^d$ such that $\varepsilon_j q_j > 2$ for all j, and $\eta \in H^r_{q,\tau}$, $\eta' \in H^{r'}_{q,\tau}$, the following equality holds:

(3.61) $$\int_{Y_\xi} \vartheta(\nu)\, d\sigma_{\chi,\xi}^{r,r'}(\nu)$$
$$= (2\pi)^{-d} |D_F|^{1/2} \alpha(\chi,\xi;r,r') \int_{Y_\xi} \vartheta(\nu)\, d\mathrm{Pl}_\xi(\nu)$$
$$+ K_\chi^{r,r'}\left(B_\xi^{\varepsilon,\varepsilon'} \vartheta\right),$$

with absolute convergence of all integrals and sums.

3. DERIVATION OF THE SPECTRAL SUM FORMULA

See (3.43) and (3.49) for the measures $d\sigma^{r,r'}_{\chi,\xi}$ and $d\text{Pl}_\xi$ on Y_ξ, (3.48) for the factor α, (3.60) for the sum of Kloosterman sums $K^{r,r'}_{\chi'}$, and (3.53) for the Bessel transform $B^{\varepsilon,\varepsilon'}_\xi$.

The function ϑ has symmetries when a coordinate v_j with $|\operatorname{Re} v_j| \leq \tau$ is multiplied by -1:

(3.62) $$\vartheta\left(v^{(j)}\right) = (\varepsilon_j \varepsilon'_j)^{\xi_j} \vartheta(v) \qquad \text{for } j = 1, \ldots, d,$$

where $v^{(j)} \in \mathbb{C}^d$ has coordinates $\left(v^{(j)}\right)_j = -v_j$, $\left(v^{(j)}\right)_l = v_l$ for $l \neq j$.

In the formula (3.61), we need only the values of ϑ on $Y_\xi \subset (i\mathbb{R} \cup [0, \infty))$, see (3.42). However, the ϑ are defined on the larger set $D(\xi, \tau)$, see (3.6) and (3.7), and depend holomorphically on the coordinates with $|\operatorname{Re} v_j| \leq \tau$. If $\varepsilon_j \varepsilon'_j = -1$, the gamma factors introduce zeros for $v_j \in \frac{\xi_j - 1}{2} + \mathbb{Z}$. If $\varepsilon_j \varepsilon'_j = 1$, the gamma factors do not lead to zeros with $|\operatorname{Re} v_j| \leq \tau$.

3.4. Kloosterman term. The test functions ϑ in Proposition 3.9 form a class of functions that is awkward to use in practice. We want to be able to pick ϑ freely, and not to construct it from other test functions η, η' and gamma factors. Our aim is to prove the sum formula for a larger class of test functions. For the extension process, we need the convergence of the delta term and the Kloosterman term for functions ϑ for which we do not yet know that the sum formula holds. The delta term is relatively simple. In this subsection we consider the convergence of the Kloosterman term.

3.4.1. *Bessel transform.* Estimates of the Bessel transforms are essential to prove the convergence of the Kloosterman term without use of the sum formula. We keep the choices and notations of §3.3.1.

The power series expansion (3.27) implies

(3.63) $$J^{\pm}_{2\nu}(y) \ll_{y_0} \frac{y^{2\operatorname{Re}\nu}}{|\Gamma(2\nu + 1)|} \qquad \text{for } \operatorname{Re}\nu \geq -\frac{1}{4},\ 0 < y \leq y_0,$$

for each $y_0 > 0$. This shows that (3.53) converges for each t if the factors φ_j of φ satisfy $\varphi_j(v) \ll (1 + |v|)^{-a}$ for $\operatorname{Re} v = 0$, with $a > \frac{3}{2}$, and if $\varphi_j\left(\frac{\xi_j + 1}{2} + h\right)$ has polynomial growth in $h \in \mathbb{N}$. We shall also want $\varphi = \bigotimes_j \varphi_j$ to be integrable for the measure Pl_ξ. So it seems sensible to use *test functions* of product form, with factors φ_j in the following class:

DEFINITION 3.10. Let $\xi \in \{0, 1\}$, $\tau > 0$, $\tau \notin \frac{1}{2}\mathbb{Z}$, and $a > 2$. The class $T^{\pm 1}_\xi(\tau, a)$ consists of the functions φ on

$$D(\xi, \tau) = \{v \in \mathbb{C} : |\operatorname{Re} v| \leq \tau\} \cup \left(\frac{\xi + 1}{2} + \mathbb{N}_0\right),$$

satisfying the following conditions:
  (T1) φ is holomorphic on the strip $|\operatorname{Re} v| \leq \tau$;
  (T2) $\varphi(v) \ll (1 + |v|)^{-a}$ for $v \in D(\xi, \tau)$;
  (T3) $\varphi(-v) = (\pm 1)^\xi \varphi(v)$ on the strip $|\operatorname{Re} v| \leq \tau$;
  (T4) If $\pm 1 = -1$, then $\varphi\left(\frac{b-1}{2}\right) = 0$ for all $b \equiv \xi \bmod 2$, $b > 1 - 2\tau$.

This leads to the following class of *principal test functions*:

DEFINITION 3.11. Let $\xi \in \{0, 1\}^d$, $\tau > 0$, $\tau \notin \frac{1}{2}\mathbb{Z}$. The class $T^{\varepsilon\varepsilon'}_\xi(\tau, a)$ consists of the functions on $D(\xi, \tau) = \prod_{j=1}^d D(\xi_j, \tau)$ of the form $\varphi(v) = \prod_{j=1}^d \varphi_j(v_j)$ with $\varphi_j \in T^{\varepsilon_j \varepsilon'_j}_{\xi_j}(\tau, a)$.

The functions $\vartheta^{r,r'}_q$ in (3.45) belong to $T^{\varepsilon\varepsilon'}_\xi(\tau, a)$.

To find properties of the Bessel transform, we first work locally, employing integral representations of Bessel functions. For the moment, $\xi \in \{0, 1/2\}$, $\eta, \eta' \in \{1, -1\}$. For

$\varphi \in T_\xi^{\eta\eta'}(\tau, a)$, the local Bessel transform $B_\xi^{\eta,\eta'} \varphi$ in (3.54) converges absolutely, and can be rewritten as follows:

$$(3.64) \quad (B_\xi^{\eta,\eta'} \varphi)(t) = 2\pi (i\eta \operatorname{sign} t)^\xi \int_{\operatorname{Re} v = \alpha} \varphi(v) \frac{v J_{2v}^{\eta\eta'}(|t|)}{\cos \pi(v - \frac{\xi}{2})} \frac{dv}{2\pi i}$$

$$+ \sum_{b > 2\alpha + 1,\, b \equiv \xi \bmod 2} (-1)^{(b-\xi)/2} \varphi_j\left(\tfrac{b-1}{2}\right) (b-1)(i\eta \operatorname{sign} t)^{\xi_j} J_{b-1}(|t|),$$

for each $\alpha \in [0, \tau]$, $\alpha \notin \tfrac{1}{2}\mathbb{Z}$. If $\eta\eta' = -1$, the sum over b vanishes, and

$$(3.65) \quad (B_\xi^{\eta,\eta'} \varphi)(t) = -\int_{\operatorname{Re} v = \alpha} \varphi(n) 4v (i\eta \operatorname{sign} t)^\xi \sin \pi\left(v + \tfrac{\xi}{2}\right) K_{2v}(|t|) \frac{dv}{2\pi i}$$

with $0 \leq \alpha \leq \tau$, $\alpha < \tfrac{1}{2}$.

The integral representation (3.64) applied with $\alpha = \tau$ and estimate (3.63), imply that $(B_\xi^{\eta,\eta'} \varphi)(t) \ll |t|^{2\tau_1 + 1}$ for $0 < |t| \leq 1$. An estimate $(B_\xi^{\eta,\eta'} \varphi)(t) = O(1)$ for $|t| \geq 1$ follows from the following integral representations of Bessel functions:

$$(3.66) \quad J_v(y) = \frac{1}{\pi} \int_0^{\pi/2} \cos(v\vartheta - y \sin \vartheta)\, d\vartheta$$

$$+ \frac{1}{\pi} \int_0^\infty e^{-vt} \sin\left(y \cosh t - \tfrac{\pi v}{2}\right) dt \quad (\operatorname{Re} v > 0),$$

$$(3.67) \quad K_{2v}(y) = \frac{1}{\sqrt{\pi}} \Gamma\left(2v + \tfrac{1}{2}\right) 2^{2v-1} y^{-2v}$$

$$\cdot \int_{-\infty}^\infty e^{-iyu} \left(1 + u^2\right)^{-2v - 1/2} du \quad (\operatorname{Re} v > 0).$$

See (7) in §6.2 of [50] for (3.66), and Basset's integral 6.16 for (3.67). Thus, we obtain $J_{2v}(y) \ll e^{\pi |\operatorname{Im} v|}$ uniformly in $y > 0$ and $\operatorname{Re} v \geq \tfrac{1}{4}$. Take $\alpha \in \left(0, \tfrac{1}{2}\right)$ for the application of (3.66). For the other case, use $K_{2v}(y) \ll_\alpha (1 + |\operatorname{Im} v|)^{2\alpha} e^{-\pi |\operatorname{Im} v|} y^{-\alpha}$ uniformly in $y > 0$ and $\operatorname{Re} v = \alpha \in \left(0, \tfrac{a}{2} - 1\right)$.

LEMMA 3.12. *Let $\tau > 0$, $\xi \in \{0, 1\}$, $\varepsilon, \varepsilon' \in \{1, -1\}$. Let $\alpha > 0$, $\delta > 0$, such that $\delta < 2\tau + a - \tfrac{3}{2}$, $\delta < a - 1$, $2\alpha + \delta < a - 2$. For $\varphi \in T_\xi^{\varepsilon\varepsilon'}(\tau, a)$, put*

$$C_{\tau,a,\alpha,\delta}(\varphi) = \sup_{\operatorname{Re} v = \tau} |v|^{3/2 - 2\tau + \delta} |\varphi(v)|$$

$$+ (\text{if } \varepsilon\varepsilon' = -1) \sup_{\operatorname{Re} v = \alpha} |v|^{2 + 2\alpha + \delta} |\varphi(v)|$$

$$+ (\text{if } \varepsilon\varepsilon' = 1) \sup_{b \equiv \xi \bmod 2,\, b > 1} b^{1+\delta} \left|\varphi\left(\tfrac{b-1}{2}\right)\right|.$$

Then the integral representation of $B_\xi^{\varepsilon,\varepsilon'} \varphi(t)$ in (3.54) converges absolutely. The estimate

$$(3.68) \quad B_\xi^{\varepsilon,\varepsilon'} \varphi(t) \ll_{\tau,a,\alpha,\delta} C_{\tau,a,\alpha,\delta}(\varphi) \min\left(|t|^{2\tau}, 1\right)$$

holds uniformly in $t \in \mathbb{R}^*$ and $\varphi \in T_\xi^{\varepsilon\varepsilon'}(\tau, a)$.

PROOF. The conditions on α and δ imply that $C_{\tau,a,\alpha,\delta}(\varphi) < \infty$, see (T2). For $|t| \leq 1$, we use (3.63) and (3.64), and for $t \geq 1$ either (3.66) or (3.67). □

These local considerations imply the following result:

LEMMA 3.13. *Let $\tau > \frac{1}{2}$, and $\varphi \in T_\xi^{\varepsilon\varepsilon'}(\tau, a)$. The integral (3.53) converges absolutely,*

$$B_\xi^{\varepsilon,\varepsilon'}\varphi(t) \ll \prod_{j=1}^d \min\left(|t_j|^{2\tau}, 1\right),$$

and $B_\xi^{\varepsilon,\varepsilon'}\varphi$ is a continuous function on $(\mathbb{R}^*)^d$ satisfying

(3.69) $$\left(B_\xi^{\varepsilon,\varepsilon'}\varphi\right)(t) \ll_{a,\tau} \prod_{j=1}^d \min\left(|t_j|^{2\tau}, 1\right).$$

3.4.2. Convergence of sums of Kloosterman sums. We consider the absolute convergence of $K_\chi^{r,r'}(f)$, see (3.60), for functions on $(\mathbb{R}^*)^d$ that satisfy an estimate

(3.70) $$|f(t)| \le C\prod_{j=1}^d \min\left(|t_j|^\alpha, |t_j|^{-\beta}\right),$$

for some $\alpha, \beta \in \mathbb{R}$, $C \ge 0$. In the course of the computation, we shall see what conditions to put on α and β.

We use an estimate $S_\chi(r', r; c) \ll_b |N(A)| \, |N(B)|^b$ for some $b \in \mathbb{R}$, where $(c) = AB$ with B an ideal in I relatively prime to I, and where A is a product of powers of prime ideals dividing I. A trivial estimate gives $b = 1$ (with $\le$ instead of $\ll$). The Weil bound (2.47) implies that any $b > \frac{1}{2}$ can be used.

The sum over $c \in I \setminus \{0\}$ in (3.60) can be split up according to $c = \zeta c_0$, with $\zeta \in O^*$, and c_0 running over representatives of non-zero principal ideals in O. We write $(c_0) = AB$ as above.

To apply Lemma 2.2 to the sum over the units ζ, we have to assume that $\alpha + \beta > 0$. For each small $\delta > 0$:

(3.71) $$\sum_{\zeta \in O^*} \left|\frac{S_\chi(r', r; \zeta c_0)}{|N(\zeta c_0)|} f\left(\frac{4\pi|rr'|^{1/2}}{\zeta c_0}\right)\right| \ll_{b,\delta} CN(A)^\delta N(B)^{b-1+\delta}$$
$$\cdot |N(rr')|^{\delta/2} \min\left(\frac{|N(rr')|^{\alpha/2}}{N(A)^\alpha N(B)^\alpha}, \frac{N(A)^\beta N(B)^\beta}{|N(rr')|^{\beta/2}}\right).$$

The important point is the absolute convergence of the sum of Kloosterman sums. If we were looking for an estimate, we would split up the sum at $N(A)N(B) = |N(rr')|^{1/2}$. At present, we note that the first option under the minimum gives

(3.72) $$\sum_{c \in I}{}' \left|\frac{S_\chi(r', r; c)}{|N(c)|} f\left(\frac{4\pi|rr'|^{1/2}}{c}\right)\right|$$
$$\ll_{b,\delta} C|N(rr')|^{(\alpha+\delta)/2} \sum_{A,B} N(A)^{\delta-\alpha} N(B)^{b+\delta-\alpha-1}$$
$$\ll C|N(rr')|^{(\alpha+\delta)/2} \prod_{P|I} \frac{1}{1 - N(P)^{\delta-\alpha}} \prod_{P\nmid I} \frac{1}{1 - N(P)^{b+\delta-\alpha-1}}$$
$$< \infty,$$

provided $\alpha > \delta$ and $\alpha > b + \delta$. The positive δ can be chosen as small as we like. With the trivial estimate of Kloosterman sums ($b = 1$), we need $\alpha > 1$. With the Weil type estimate (2.47) ($b > \frac{1}{2}$), we obtain:

PROPOSITION 3.14. *If the function f on $(\mathbb{R}^*)^d$ satisfies estimate (3.70) with $\alpha > \frac{1}{2}$ and $\alpha + \beta > 0$, then the sum of Kloosterman sums $\mathrm{K}_\chi^{r,r'}(f)$ converges absolutely, and is estimated by*

$$\sup_{t \in (\mathbb{R}^*)^d} \frac{|f(t)|}{\prod_{j=1}^d \min\left(|t_j|^\alpha, |t_j|^{-\beta}\right)}.$$

COROLLARY 3.15. *If $\varphi \in T_\xi^{\varepsilon\varepsilon'}(\tau_1, a)$ with $\frac{1}{4} < \tau_1 < \frac{1}{2}$, then the sum of Kloosterman sums $\mathrm{K}_\chi^{r,r'}\left(\mathrm{B}_\xi^{\varepsilon,\varepsilon'}\varphi\right)$ converges absolutely.*

3.5. Extension. The extension of the restricted sum formula (3.61) to a larger class of test functions requires a limiting process. The basic idea is present in §16.3 of [**4**]. There, and in [**9**], Γ is allowed to be more general than here, and a Weil-Salié type estimate of Kloosterman sums is not available. That leads to a sum formula for test functions on a wide strip, $|\operatorname{Re}\nu| \leq \tau$ with $\tau > \frac{1}{2}$. With the Weil-Salié estimate, test functions on a narrow strip, $\frac{1}{4} < \tau < \frac{1}{2}$, can be used.

We make the method explicit by formulating first two extension lemmas. For (products of) $\mathrm{SL}_2(\mathbb{R})$ this is the first such formulation as far as we know. Similar extension lemmas for $\mathrm{SL}_2(\mathbb{C})$ are used implicitly in [**10**] and stated explicitly in [**33**].

3.5.1. *Extension lemmas.* If ϑ is a function on Y, see (3.42), we shall say that *the sum formula holds for (r, r', ϑ)*, if equality (3.61) holds, with absolute convergence of all sums and integrals.

Up till now, we have used test functions in $T_\xi^{\varepsilon\varepsilon'}(\tau, a)$ with $\tau > \frac{1}{2}$. We shall extend the class of test functions to $T_\xi^{\varepsilon\varepsilon'}(\tau_1, a)$ with $\tau_1 \in \left(\frac{1}{4}, \frac{1}{2}\right)$. Until we have reached the final version of the spectral sum formula, in Theorem 3.21, we shall denote by τ a number larger than $\frac{1}{2}$, and by τ_1 a number larger than $\frac{1}{4}$.

To extend the class of functions for which the sum formula holds, we use two lemmas. The first is based on Fatou's lemma, and has a bootstrap character: It really extends the sum formula to a larger class of test functions.

LEMMA 3.16. *Let $r \in O'$, $r \neq 0$, and let $\tau_1 > \frac{1}{4}$ satisfy assumption (3.78). Suppose that the function φ and the sequence of functions (φ_n) on Y_ξ satisfy the following conditions:*

i) *The sum formula holds for (r, r, φ_n) for each n.*
ii) *The integral defining $\mathrm{B}_\xi^{\varepsilon,\varepsilon}\varphi$ converges absolutely, and*

$$\lim_{n \to \infty} \left(\mathrm{B}_\xi^{\varepsilon,\varepsilon}\varphi_n\right)(t)/\mathrm{m}(\tau_1, t) = \left(\mathrm{B}_\xi^{\varepsilon,\varepsilon}\varphi\right)(t)/\mathrm{m}(\tau_1, t)$$

uniformly in $t \in (\mathbb{R}^*)^d$, with $\mathrm{m}(\tau_1, t) := \prod_j \min\left(|t_j|^{2\tau_1}, 1\right)$.
iii) *φ is integrable for $d\mathrm{Pl}_\xi$.*
iv) *$\varphi_n(\nu) \geq 0$ and $\varphi_n(\nu) \to \varphi(\nu)$ for each $\nu \in Y_\xi^{\tau_1}$.*

Then the sum formula holds for (r, r, φ), and

(3.73) $$\lim_{n \to \infty} \mathrm{K}_\chi^{r,r}\left(\mathrm{B}_\xi^{\varepsilon,\varepsilon}\varphi_n\right) = \mathrm{K}_\chi^{r,r}\left(\mathrm{B}_\xi^{\varepsilon,\varepsilon}\varphi\right),$$

(3.74) $$\lim_{n \to \infty} \int_{Y_\xi} \varphi_n(\nu) \, d\sigma_{\chi,\xi}^{r,r}(\nu) = \int_{Y_\xi} \varphi(\nu) \, d\sigma_{\chi,\xi}^{r,r}(\nu),$$

(3.75) $$\lim_{n \to \infty} \int_{Y_\xi} \varphi_n(\nu) \, d\mathrm{Pl}_\xi(\nu) = \int_{Y_\xi} \varphi(\nu) \, d\mathrm{Pl}_\xi(\nu).$$

The second lemma is based on Lebesgue's theorem on dominated convergence:

3. DERIVATION OF THE SPECTRAL SUM FORMULA

LEMMA 3.17. *Let $r, r' \in O' \setminus \{0\}$, and suppose that $\tau_1 > \frac{1}{4}$ satisfies (3.78). Suppose that the functions φ, η, and the sequence of functions (φ_n) on Y_ξ satisfy the following conditions:*
  a) *The sum formula holds for (r, r', φ_n) for each n.*
  b) *The sum formula holds for (r, r, η) and for (r', r', η).*
  c) *The integral defining $\mathrm{B}_\xi^{\varepsilon,\varepsilon'} \varphi$ converges absolutely, there exists $C \geq 0$ such that*
  $$\left|\left(\mathrm{B}_\xi^{\varepsilon,\varepsilon'} \varphi_n\right)(t)\right| \leq C\, \mathrm{m}(\tau_1, t) \text{ for all } t \in (\mathbb{R}^*)^d \text{ and for all } n, \text{ and}$$
  $$\lim_{n \to \infty} \left(\mathrm{B}_\xi^{\varepsilon,\varepsilon'} \varphi_n\right)(t) = \left(\mathrm{B}_\xi^{\varepsilon,\varepsilon'} \varphi\right)(t) \qquad \text{for each } t \in (\mathbb{R}^*)^d.$$
  d) $\varphi_n \to \varphi$ *pointwise on Y_ξ.*
  e) $|\varphi_n| \leq \eta$ *on Y_ξ.*

Then the sum formula holds for (r, r', φ), and

(3.76) $$\lim_{n \to \infty} \mathrm{K}_\chi^{r,r'}\left(\mathrm{B}_\xi^{\varepsilon,\varepsilon'} \varphi_n\right) = \mathrm{K}_\chi^{r,r'}\left(\mathrm{B}_\xi^{\varepsilon,\varepsilon'} \varphi\right),$$

(3.77) $$\lim_{n \to \infty} \int_{Y_\xi} \varphi_n(v)\, d\sigma_{\chi,\xi}^{r,r'}(v) = \int_{Y_\xi} \varphi(v)\, d\sigma_{\chi,\xi}^{r,r'}(v),$$

and (3.75) are satisfied.

PROOF OF LEMMA 3.16. Let us first look at the Kloosterman term. For each $\delta > 0$ we have for all sufficiently large n and for all $c \in I \setminus \{0\}$:

$$\left|\left(\mathrm{B}_\xi^{\varepsilon,\varepsilon} \varphi_n\right)\left(\tfrac{4\pi|r|}{c}\right) - \left(\mathrm{B}_\xi^{\varepsilon,\varepsilon} \varphi\right)\left(\tfrac{4\pi|r|}{c}\right)\right| \leq \delta\, \mathrm{m}\left(\tau_1, 4\pi|rr'|^{1/2} c^{-1}\right).$$

Proposition 3.14 gives the absolute convergence of

$$\sideset{}{'}\sum_{c \in I} \mathrm{m}\left(\tau_1, \tfrac{4\pi|rr'|^{1/2}}{c}\right).$$

From

$$\left|\mathrm{B}_\xi^{\varepsilon,\varepsilon} \varphi(t)\right| \leq \left|\mathrm{B}_\xi^{\varepsilon,\varepsilon} \varphi_n(t)\right| + \left|\mathrm{B}_\xi^{\varepsilon,\varepsilon} \varphi(t) - \mathrm{B}_\xi^{\varepsilon,\varepsilon} \varphi_n(t)\right|,$$

we conclude that $\mathrm{K}_\chi^{r,r}\left(\mathrm{B}_\xi^{\varepsilon,\varepsilon} \varphi\right)$ converges absolutely. Letting $\delta \downarrow 0$ gives (3.73). Conditions iii) and iv) imply (3.75), by dominated convergence.

As the sum formula holds for (r, r, φ_n), we have a pointwise convergent sequence of $d\sigma_{\chi,\xi}^{r,r}$-integrable functions on Y_ξ for which $\lim_{n \to \infty} \int_{Y_\xi} \varphi_n(v)\, d\sigma_{\chi,\xi}^{r,r}(v)$ exists. So the limit function φ is integrable for $d\sigma_{\chi,\xi}^{r,r}$ by Fatou's lemma. Dominated convergence gives (3.74). $\square$

PROOF OF LEMMA 3.17. The sums $\mathrm{K}_\chi^{r,r'}\left(\mathrm{B}_\xi^{\varepsilon,\varepsilon'} \varphi_n\right)$ can be estimated by

$$\sideset{}{'}\sum_{c \in I} |N(c)|^{-1/2+\delta}\, \mathrm{m}\left(\tau_1, \tfrac{4\pi|rr'|^{1/2}}{c}\right),$$

uniformly in n. So Proposition 3.14 gives the absolute convergence, and condition c) implies (3.76).

The function η is integrable for $d\mathrm{Pl}_\xi$. It can be used as a majorant to conclude that φ is integrable for $d\mathrm{Pl}_\xi$ and that (3.75) holds.

Inequality (3.44) shows that η is integrable for $\left|d\sigma_{\chi,\xi}^{r,r'}\right|$ as well. It serves as a majorant to conclude that φ is integrable for $d\sigma_{\chi,\xi}^{r,r'}$ and gives (3.77). $\square$

Remark 1. These extension lemmas are based on Proposition 3.14, and hence depend on the Weil type estimate (2.47), and on the regular spacing of the elements of $I \subset O$. The trivial estimate $|S_\chi(r, r'; c)| \leq |N(c)|$ works as well, provided that we use $\tau_1 > \frac{1}{2}$.

Remark 2. For nonarithmetic groups Γ, in the case $d = 1$, it seems hard to establish absolute convergence of the sum of Kloosterman sums $K_\chi^{r,r'}\left(B_\xi^{\varepsilon,\varepsilon'}\varphi\right)$ for test functions φ for which we do not yet have derived the sum formula.

Remark 3. The functions φ_n, φ and η in the Lemmas 3.16 and 3.17 are defined on Y_ξ, where we note that $\left(0, \frac{1}{2}\right]$ is contained in Y_{ξ_j} if $\xi_j = 0$.

When we know that there is no ϖ with $v_{\varpi,j} \in \left[\tau_1, \frac{1}{2}\right)$ for some j, then the support of the measures $d\sigma_{\chi,\xi}^{r,r'}$ is contained in

(3.78) $$\operatorname{Supp} d\sigma_{\chi,\xi}^{r,r'} \subset Y_\xi^{\tau_1} := \prod_{j, \xi_j = 1} (i[0, \infty] \cup \mathbb{N})$$
$$\times \prod_{j, \xi_j = 0} \left(i[0, \infty) \cup (0, \tau_1] \cup \left(\tfrac{1}{2} + \mathbb{N}_0\right)\right).$$

Under this condition, the statements of the extension lemmas are valid with Y_ξ replaced by $Y_\xi^{\tau_1}$.

3.5.2. *Extensions*. We start with the formula in Proposition 3.9, and apply the extension lemmas in §3.5.1 to enlarge the class of test functions in the formula.

LEMMA 3.18. *Let* $\varepsilon \in \{0, 1\}^d$, $\tau \in \left(\frac{1}{2}, 1\right)$, $p > \tau$, *and* $a > 2$. *We define the function* $\varphi_{a,p} = \bigotimes \varphi_{a,p,j}$ *by*

(3.79) $$\varphi_{a,p,j}(v) := \begin{cases} (p^2 - v^2)^{-a/2} & \text{if } |\operatorname{Re} v| \leq \tau, \\ (p^2 + v^2)^{-a/2} & \text{if } v \in \frac{\xi_j - 1}{2} + \mathbb{N}_0,\ v > \tau. \end{cases}$$

The sum formula holds for $(r, r, \varphi_{a,p})$ *for each* $r \in O' \setminus \{0\}$.

Note that $\varphi_{a,p} \in T_\xi^{\varepsilon^2}(\tau, a)$, and that each $\varphi \in T_\xi^{\varepsilon^2}(\tau, a)$ satisfies $\varphi(v) \ll_\varphi \varphi_{a,p}(v)$ on its domain.

PROOF. Let $r \in O'$, $r \neq 0$. We prove that the sum formula holds for $(r, r, \varphi_{a,p})$ by repeated application of Lemma 3.16.

For $m = (m_1, \ldots, m_d)$, with $m_j \in \mathbb{Z}_{\geq 2}$, we put $q(m)_j = \varepsilon_j\left(\xi_j + 2m_j\right)$, and

(3.80) $$\eta_{m_j, j}(v) = \begin{cases} e^{v^2/m_j}(p^2 - v^2)^{-a/2} & \text{for } |\operatorname{Re} v| \leq \tau, \\ (p^2 + v^2)^{-a/2} & \text{for } v \in \frac{\xi_j + 1}{2} + \mathbb{Z} \text{ and } 1 + 2\tau \\ & \qquad < 2|v| + 1 \leq |q(m)_j|, \\ 0 & \text{elsewhere in } \frac{\xi_j + 1}{2} + \mathbb{Z}; \end{cases}$$

$$\eta'_{m_j, j}(v) = \Gamma\left(\tfrac{1}{2} + v + \tfrac{|q(m)_j|}{2}\right)\Gamma\left(\tfrac{1}{2} - v + \tfrac{|q(m)_j|}{2}\right)$$
$$\cdot \begin{cases} e^{v^2/m_j} & \text{for } |\operatorname{Re} v| \leq \tau, \\ 1 & \text{for } v \in \frac{\xi_j + 1}{2} + \mathbb{Z}, \\ & \qquad \text{and } 1 + 2\tau < 2|v| + 1 \leq |q(m)_j|, \\ 0 & \text{elsewhere in } \frac{\xi_j + 1}{2} + \mathbb{Z}; \end{cases}$$

3. DERIVATION OF THE SPECTRAL SUM FORMULA

$$\vartheta_{m_j,j}(\nu) = \frac{\eta_{m_j,j}(\nu)\overline{\eta'_{m_j,j}(\bar{\nu})}}{\Gamma\left(\frac{1}{2} + \nu + \frac{|q(m)_j|}{2}\right)\Gamma\left(\frac{1}{2} - \nu + \frac{|q(m)_j|}{2}\right)}$$

$$= \begin{cases} e^{2\nu^2/m_j}\varphi_{a,p,j}(\nu) & \text{if } |\operatorname{Re}\nu| \leq \tau, \\ \varphi_{a,p,j}(\nu) & \text{if } |\operatorname{Re}\nu| > \tau,\ 2|\nu| + 1 \leq |q(m)_j|, \\ 0 & \text{otherwise};\end{cases}$$

$$\vartheta^{r,r}_{q(m)}(\nu) = \prod_{j=1}^{d} \vartheta_{m_j,j}(\nu).$$

We apply Proposition 3.9 to $\vartheta^{r,r}_{q(m)}$, and conclude that the sum formula holds for $(r, r, \vartheta^{r,r}_{q(m)})$.

Let us now apply Lemma 3.16 with

$$\varphi = \varphi_{a,p,1} \otimes \bigotimes_{j=2}^{d} \vartheta_{m_j,j}, \qquad \varphi_n = \vartheta_{n,1} \otimes \bigotimes_{j=2}^{d} \vartheta_{m_j,j}.$$

Conditions i) and iii) in Lemma 3.16 are satisfied, as well as $\varphi_n(\nu) \geq 0$ on Y_ξ, which is part of condition iv). We note that:

$$\varphi_n(\nu) - \varphi(\nu)$$
$$= \begin{cases} \left(e^{2\nu_1^2/n} - 1\right)\varphi_{a,p,1}(\nu_1)\prod_{j=2}^{d}\vartheta_{m_j,j}(\nu_j) & \text{for } |\operatorname{Re}\nu_1| \leq \tau, \\ 0 & \text{if } \tau < \nu_1 \leq \frac{|q(m)_1|-1}{2}, \\ -\varphi_{a,p,1}(\nu_1)\prod_{j=2}^{d}\vartheta_{m_j,j}(\nu_j) & \text{for } \nu_1 > \frac{|q(m)_1|-1}{2}.\end{cases}$$

This implies that $\varphi_n \to \varphi$ pointwise on Y_ξ, which is the other half of condition iv).

To check condition ii), we use Lemma 3.12. For $\nu_1 = \tau + iy$, we have

$$(1+|\nu_1|)^{\frac{3}{2}-2\tau+\delta}\left(e^{2\nu_1^2/n} - 1\right)\varphi_{a,p,1}(\nu_1)$$
$$\ll (1+|y|)^{\frac{3}{2}-2\tau+\delta-a}\begin{cases}(1+|y|)^2/m & \text{if } |y| \leq \sqrt{n}, \\ 1 & \text{if } |y| \geq \sqrt{n}.\end{cases}$$

Hence

$$\sup_{\operatorname{Re}\nu_1=\tau}(1+|\nu_1|)^{\frac{3}{2}-2\tau+\delta}\left|\left(e^{2\nu_1^2/n}-1\right)\varphi_{a,p,1}(\nu_1)\right|$$
$$\ll \min\left(n^{-1}, n^{\frac{3}{4}-\tau+\frac{\delta}{2}-\frac{a}{2}}, n^{\frac{3}{4}-\tau+\frac{\delta}{2}-\frac{a}{2}}\right),$$

which is $o(1)$ as $n \to \infty$, for a suitable choice of δ.

The other contributions to $C_{\tau,a,\alpha,\delta}(\varphi_{n,1} - \varphi_{a,p,1})$, see Lemma 3.12, are shown to be $o(1)$, by this or an easier method. With Lemma 3.13, condition ii) follows. We conclude that the sum formula holds for (r, r, φ).

Now we can proceed, successively applying Lemma 3.16 with

$$\varphi = \bigotimes_{j=1}^{l}\varphi_{a,p,j} \otimes \bigotimes_{j=l+1}^{d}\vartheta_{m_j,j}, \qquad \varphi_n = \bigotimes_{j=1}^{l-1}\varphi_{a,p,j} \otimes \vartheta_{n,l} \otimes \bigotimes_{j=l+1}^{d}\vartheta_{m_j,j}.$$

We conclude in the end that the sum formula holds for $(r, r, \varphi_{a,p})$. □

We now formulate the main result of this extension step as a proposition. The proof can be adapted to hold with the trivial estimate $|S_\chi(r', r; c)| \leq |N(c)|$.

PROPOSITION 3.19. *The sum formula holds for (r, r', φ) for all $r, r' \in \mathcal{O}' \setminus \{0\}$, and all $\varphi \in T_\xi^{\varepsilon\varepsilon'}(\tau, a)$ with $\tau \in \left(\frac{1}{2}, 1\right)$ and $a > 2$.*

PROOF. For a given $\varphi = \bigotimes_j \varphi_j \in T_\xi^{\varepsilon\varepsilon'}(\tau, a)$, we construct $\varphi_n = \bigotimes_j \varphi_{n,j}$ to be used in Lemma 3.17. Let

$$\psi_j(\nu) = \begin{cases} 1 & \text{if } \varepsilon_j \varepsilon_j' = 1, \\ \left(\frac{1}{4} - \nu^2\right) & \text{if } \varepsilon_j \varepsilon_j' = -1 \text{ and } \xi_j = 0, \\ \nu & \text{if } \varepsilon_j \varepsilon_j' = -1 \text{ and } \xi_j = 1. \end{cases}$$

We take

$$\eta_{n,j}(\nu) = \begin{cases} e^{\nu^2/n} \varphi_j(\nu) \frac{1}{\psi_j(\nu)} & \text{if } |\operatorname{Re}\nu| \leq \tau, \\ \varphi_j(\nu) & \text{for } \nu \in \frac{\xi_j - 1}{2} + \mathbb{N}_0 \text{ and } 1 + 2\tau \\ & < 2|\nu| + 1 \leq |q(n)_j|, \\ 0 & \text{elsewhere in } \frac{\xi_j + 1}{2} + \mathbb{Z} \end{cases}$$

In the case $\varepsilon_j \varepsilon_j' = -1$, $\xi_j = 1$, the factor ν in $\psi_j(\nu)$ ensures that η_j is even. We check that $\eta = \bigotimes_j \eta_j$ is in $H_{q,\tau}^r$. In the definition of η', we have to compensate for the gamma factors in (3.45):

$$\eta'_{n,j}(\nu) = \Gamma\left(\tfrac{1}{2} + \nu + \tfrac{|q(n)_j|}{2}\right) \Gamma\left(\tfrac{1}{2} - \nu + \tfrac{\varepsilon_j \varepsilon_j' |q(n)_j|}{2}\right)$$

$$\cdot \begin{cases} e^{\nu^2/n} \psi_j(n) & \text{if } |\operatorname{Re}\nu| \leq \tau, \\ 1 & \text{if } \varepsilon_j \varepsilon_j' = 1, \nu \in \frac{\xi_j - 1}{2} + \mathbb{N}_0, \\ & \text{and } \tau < \nu \leq \frac{|q(n)_j| - 1}{2}, \\ 0 & \text{otherwise.} \end{cases}$$

In the strip $|\operatorname{Re}\nu| \leq \tau$, the factor $\psi_j(\nu)$ compensates for poles that may come from the second gamma factor. If $\varepsilon_j \varepsilon_j' = 1$ and $\nu = \frac{b-1}{2} > \tau$, then the gamma factors are finite if $b < |q(n)_j|$. If $\varepsilon_j \varepsilon_j' = -1$ and $\nu = \frac{b-1}{2} > \tau$, then $\varphi_j(\nu) = 0$ anyhow.

We put

(3.81) $$\varphi_n(\nu) = \vartheta_{(q(n))}^{r,r'}(\nu)$$

$$= \prod_j \frac{\eta_{n,j}(\nu_j) \overline{\eta'_{n,j}(\nu_j)}}{\Gamma\left(\tfrac{1}{2} + \nu_j + \tfrac{|q(n)_j|}{2}\right) \Gamma\left(\tfrac{1}{2} - \nu + \tfrac{\varepsilon_j \varepsilon_j' |q(n)_j|}{2}\right)},$$

as in (3.45). Proposition 3.9 shows that condition a) in Lemma 3.17 is satisfied.

Choose $p > \tau$. We use $C\varphi_{a,p}$ as the majorant η in Lemma 3.17, where we choose C such that condition e) holds. Condition b) follows from Lemma 3.18.

We note that

(3.82) $$\varphi_{n,j}(\nu) = \varphi_j(\nu) \cdot \begin{cases} e^{\nu^2/n} & \text{if } |\operatorname{Re}\nu| \leq \tau, \\ 1 & \text{if } \varepsilon_j \varepsilon_j' = 1, \nu \equiv \xi_j \bmod 2, \tau < \nu \leq \frac{|q(n)_j| - 1}{2}, \\ 0 & \text{otherwise.} \end{cases}$$

This gives conditions c) and d), the former by application of Lemma 3.12. □

3.5.3. *Exceptional eigenvalues.* Spectral parameters v_ϖ such that some component $v_{\varpi,j} \in \left(0, \tfrac{1}{2}\right)$ are called *exceptional*. This can only occur if $\xi_j = 0$. The corresponding eigenvalue $\lambda_\varpi = (\lambda_{\varpi,1}, \ldots, \lambda_{\varpi,d})$ is called an exceptional eigenvalue. According to the Ramanujan conjecture, there should be no exceptional eigenvalues for any congruence subgroup of $\mathrm{SL}_2(F) \subset \mathrm{SL}_2(\mathbb{R})^d$. At present, it is known that exceptional spectral parameters, if present, are confined to a small subinterval $(0, a)$ of $\left(0, \tfrac{1}{2}\right)$. The best result known to us gives $a = \tfrac{1}{9}$, Kim and Shahidi, [**24**].

The ultimate aim in this section is to prove that the sum formula holds for all $\varphi \in T^{\varepsilon\varepsilon'}_\xi(\tau_1, a)$ with $\tfrac{1}{4} < \tau_1 < \tfrac{1}{2}$.

We can proceed on the basis of the result of Kim and Shahidi, and work with Y^τ_ξ in the extension Lemmas 3.16 and 3.17; see Remark 3 at the end of §3.5.1. To keep the exposition self-contained we shall show that there are no $v_{\varpi,j} \in \left(\tfrac{1}{4}, \tfrac{1}{2}\right)$. This extends Selberg's result, [**43**], p. 13, which shows that if $d = 1$ there are no exceptional eigenvalues in $\left(\tfrac{1}{4}, \tfrac{1}{2}\right)$. See also [**6**], Proposition 7.1.

PROPOSITION 3.20. *Every exceptional spectral parameter $v_{\varpi,j}$ is contained in the interval $\left(0, \tfrac{1}{4}\right]$.*

PROOF. Suppose that ℓ is a place for which there are cuspidal automorphic representations ϖ with $\tfrac{1}{4} < v_{\varpi,\ell} < \tfrac{1}{2}$. So $\xi_\ell = 0$. We fix $\alpha > \tfrac{1}{4}$ such that $v_{\varpi,\ell} > \alpha$ for some ϖ.

We apply Proposition 3.19 with $r = r' \in O' \setminus \{0\}$ and a test function $\varphi = \otimes_j \varphi_j$ such that

$$\varphi_\ell(v) = e^{uv^2} \cos \pi v \qquad u > 0,$$

and $\varphi_j > 0$ on Y_{ξ_j} for $j \neq \ell$, say $\varphi_j = \varphi_{a,p,j}$ in (3.79).

We consider the behavior as $u \to \infty$ of the terms in the sum formula. The dependence of the delta term on u is governed by the factor

$$\int_{Y_0} \varphi_\ell \, d\mathrm{Pl}_0 = \frac{i}{u} \int_{\mathrm{Re}\, v = 0} e^{v^2} v \sin \tfrac{\pi v}{\sqrt{u}} \, dv = o\left(\tfrac{1}{u}\right) \qquad (u \to \infty).$$

The Bessel transform $B^{\varepsilon,\varepsilon}_\xi \varphi(t)$ has factors estimated by $O\left(\min\left(|t_j|^{2\alpha}, 1\right)\right)$, as we have seen in §3.4.2. To get hold of the influence of u, we first use (3.64) and (3.63) to find for $|t_\ell| \leq 1$:

$$\left(B_0^{\varepsilon_\ell, \varepsilon_\ell} \varphi_\ell\right)(t_\ell) \ll_\alpha e^{u\alpha^2} \int_{-\infty}^\infty e^{-uy^2} (1 + |y|)^{\tfrac{1}{2} - 2\alpha} e^{\tfrac{\pi}{2}|y|} \, dy \, |t_\ell|^{2\alpha}$$
$$= o\left(u^{-1/2} e^{u\alpha^2} |t_\ell|^{2\alpha}\right) \qquad (u \to \infty).$$

For $|t_\ell| \geq 1$, we use the consequence $J_{2\nu}(|t_\ell|) \ll e^{\pi|\mathrm{Im}\,\nu|}$ of (3.66) to obtain the estimate

$$\left(B_0^{\varepsilon_\ell, \varepsilon_\ell} \varphi_\ell\right)(t_\ell) = o\left(u^{-1/2} e^{u\alpha^2}\right) \qquad (u \to \infty).$$

The dependence on u can be accounted for by taking $u^{-1/2} e^{u\alpha^2}$ into the factor C in (3.70). With Proposition 3.14, we find that

$$K^{r,r}_\xi \left(B^{\varepsilon,\varepsilon}_\xi \varphi\right) = o\left(u^{-1/2} e^{u\alpha^2}\right) \qquad (u \to \infty).$$

This is larger than the estimate of the delta term. Proposition 3.19 yields the same bound for $\int_{Y_\xi} \varphi(v) \, d\sigma^{r,r}_{\chi,\xi}(v)$.

The measure $d\sigma_{\chi,\xi}^{r,r}$ is nonnegative, and $\varphi \geq 0$ on its support; see (3.43). So for each ϖ:
$$|c^r(\varpi)|^2 \varphi(v_\varpi) = o\left(u^{-1/2} e^{u\alpha^2}\right) \quad (u \to \infty).$$

Let $v_{\varpi,\ell} \in \left(\alpha, \tfrac{1}{2}\right)$. The factor $e^{uv_{\varpi,\ell}^2} \cos \pi v_{\varpi,\ell}$ is not $o\left(u^{-1/2} e^{u\alpha^2}\right)$. We have arranged that $\varphi_j(v_{\varpi,j}) > 0$ for $j \neq \ell$. Hence $c^r(\varpi) = 0$. This holds for all $r \in \mathcal{O}' \setminus \{0\}$. The cusp forms in the space of ϖ are determined by their Fourier expansion. So we have reached a contradiction. □

3.5.4. Final extension. Now we work with a test function $\varphi \in T_\xi^{\varepsilon\varepsilon'}(\tau_1, a)$, with $\tfrac{1}{4} < \tau_1 < \tfrac{1}{2}$. We have to approximate it by functions in $T_\xi^{\varepsilon\varepsilon'}(\tau, a)$ with $\tfrac{1}{2} < \tau < 1$. This calls for a convolution $\varphi * k_n$ where k_n is, say, a Gauss kernel approximating the delta distribution at 0. It seems sensible to do this for each factor. As k_n is even, the convolution behaves in the right way under $v \mapsto -v$, so condition (T3) in the definition of $T_\xi^{(1,\ldots,1)}(\tau, a)$ is preserved; see Definition 3.10. However, in factors with $\varepsilon_j \varepsilon_j' = -1$, the zeros required in condition (T4) in Definition 3.10 will be lacking. In factors with $\xi_j = 0$, $\varepsilon_j \varepsilon_j' = 1$, there is another problem: the value $\varphi_j\left(\pm \tfrac{1}{2}\right)$ is prescribed, and should be approximated by $\varphi_j * k_n\left(\pm \tfrac{1}{2}\right)$.

We take for $|\operatorname{Re} v| \leq \tau$:

$$(3.83) \quad \varphi_{n,j}(v) = \begin{cases} -i\sqrt{\tfrac{n}{\pi}}\left(v^2 - \tfrac{1}{4}\right) \int_{\operatorname{Re}\mu=\alpha} \frac{\varphi_j(\mu)}{\mu^2 - \tfrac{1}{4}} e^{n(\mu-v)^2} d\mu \\ \quad + e^{n(v^2 - \tfrac{1}{4})} \varphi_j\left(\tfrac{1}{2}\right) & \text{if } \xi_j = 0, \\ -i\sqrt{\tfrac{n}{\pi}} \int_{\operatorname{Re}\mu=\alpha} \varphi_j(\mu) e^{n(\mu-v)^2} d\mu & \text{if } \xi_j = 1, \end{cases}$$

with $|\alpha| \leq \tau_1$; for $v > \tau$, we take $\varphi_{n,j} = \varphi_j$.

On the wide strip, $\varphi_{n,j}$ is holomorphic. Most conditions in Definition 3.10 are clearly satisfied. For (T2), we note that the integral over $|\mu| \geq |v|/2$ can be estimated by

$$(1 + |v|)^{-a-2} \int_{\operatorname{Re}\mu=\alpha} |e^{n(\mu-v)^2}| \, |d\mu| \ll_n (1 + |v|)^{-a-2},$$

and the integral over the remaining part by $\int_{|\mu| \geq \frac{|v|}{2}} |e^{nv^2}| \, |d\mu|$, which is smaller.

We apply Lemma 3.17 with $\eta = \varphi_{a,p}$. Conditions a) and b) are clear. To establish condition c), we first consider for $v = iy$, $\xi_j = 0$:

$$\varphi_{n,j}(iy) \ll \left(y^2 + \tfrac{1}{4}\right) \int_{-\infty}^\infty \left(1 + |y + t/\sqrt{n}|\right)^{-a-2} e^{-t^2} dt + e^{-n(1/4+y^2)}$$

$$\ll (1 + |y|)^2 \left(\left(1 + \tfrac{|y|}{2}\right)^{-a-2} \int_{t=-y\sqrt{n}/2}^{y\sqrt{n}/2} e^{-t^2} dt + \int_{|t| \geq y\sqrt{n}/2} e^{-t^2} dt\right)$$

$$\ll (1 + |y|)^{-a},$$

uniformly in n. The contribution of $e^{n(-y^2-1/4)}$ is $O(1)$. On $v = \tfrac{b-1}{2}$, $b \geq 2$, we have $\varphi_{n,j} = \varphi_j$. For $\xi_j = 1$, we carry out a similar computation. So the local Bessel transforms $B_{\xi_j}^{\varepsilon_j, \varepsilon_j'} \varphi_{n,j}(u)$ in (3.54) converge pointwise to $B_{\xi_j}^{\varepsilon_j, \varepsilon_j'} \varphi_j(u)$. Then the same holds for $B_\xi^{\varepsilon,\varepsilon'} \varphi_n$. This gives the pointwise convergence in condition c) in Lemma 3.17.

To get the uniform estimate by $m(\tau_1, t)$, we again work locally. The case $v \in \tfrac{\xi_j - 1}{2} + \mathbb{N}_0$ poses no problem. For $|t| \leq 1$, we need to integrate over the line $\operatorname{Re} v = \tau_1$, inside the narrow strip; see Lemma 3.12. For $|t| \geq 1$, there are also integrals over $\operatorname{Re} v = \sigma$ with $0 < \sigma \leq \tau_1$. For $v = \sigma + iy$ on such a line, we have $e^{n(\sigma^2 - y^2 - 1/4)} = O(1)$, which handles the term with $\varphi_j\left(\tfrac{1}{2}\right)$, if $\xi_j = 0$. For both values of ξ_j, we take the line of integration $\alpha = \sigma$

in (3.83), and get a uniform bound by $(1 + |\nu|)^{-a}$ by the same method as used above. This leads to an uniform bound by $m(\tau_1, t)$, establishing condition c) in Lemma 3.17. We also obtain conditions d) and e).

Thus, we have proved that the sum formula holds for all test functions in the class $T_\xi^{\varepsilon\varepsilon'}(\tau, a)$.

3.6. Sum formula. We now have the first main result in this paper, the *spectral sum formula*. We state it, and, for the convenience of the reader, follow it by a recapitulation of the terms used in it.

THEOREM 3.21. *Let $r, r' \in O' \setminus \{0\}$. Let $\xi \in \{0, 1\}^d$ determine a central character compatible with χ. Put $\varepsilon = \mathrm{sign}(r)$, $\varepsilon' = \mathrm{sign}\, r'$. For each $\varphi \in T_\xi^{\varepsilon\varepsilon'}(\tau, a)$ with $\frac{1}{4} < \tau < \frac{1}{2}$ and $a > 2$, the following equality holds, with absolute convergence of all integrals and sums:*

(3.84)
$$\int_{Y_\xi} \varphi(\nu)\, d\sigma_{\chi,\xi}^{r,r'}(\nu)$$
$$= (2\pi)^{-d} |D_F|^{1/2} \alpha(\chi, \xi; r, r') \int_{Y_\xi} \varphi(\nu)\, d\mathrm{Pl}_\xi(\nu)$$
$$+ K_\chi^{r,r'}\left(B_\xi^{\varepsilon,\varepsilon'} \varphi\right).$$

We work with a totally real number field F of degree d over $\mathbb{Q}$, with ring of integers O, a non-zero ideal $I \subset O$, and a character χ of $(O/I)^*$ satisfying $\chi(-1) = e^{\pi i \sum_j \xi_j}$.

The field F is considered as embedded in $\mathbb{R}^d$ by the d embeddings $\xi \mapsto \xi_j = \xi^{\sigma_j}$. The "Fourier term orders" r and r' are elements of the complementary ideal $O' \subset F$. The sign is to be understood as $\varepsilon = (\mathrm{sign}(r_1), \ldots, \mathrm{sign}(r_d))$. The discriminant of F over $\mathbb{Q}$ is denoted D_F.

We use *test functions* of product form $\varphi(\nu_1, \ldots, \nu_d) = \prod_{j=1}^d \varphi_j(\nu_j)$. The domain of the j-th factor is
$$\left\{\nu_j \in \mathbb{C} : |\mathrm{Re}\,\nu_j| \leq \tau\right\} \cup \left(\tfrac{\xi_j+1}{2} + \mathbb{N}_0\right).$$

Definition 3.10 requires that φ_j is holomorphic on the strip $|\mathrm{Re}\,\nu_j| \leq \tau$, and satisfies $\varphi_j(-\nu_j) = (\varepsilon_j \varepsilon_j')^{\xi_j} \varphi_j(\nu_j)$ on this strip, with $\varepsilon_j = \mathrm{sign}(r_j)$. This means that φ_j is odd if $\varepsilon_j \varepsilon_j' = -1$ and $\xi_j = 1$, and that it is even otherwise. If $\varepsilon_j \varepsilon_j' = -1$ and $\xi_j = 0$, then φ_j should vanish at $\pm\frac{1}{2}$. On the whole domain, the estimate $\varphi_j(\nu_j) \ll (1 + |\nu_j|)^{-a}$ with $a > 2$ is required.

The domain of φ contains the set
$$Y_\xi = \prod_{j,\xi_j=0} \left(i\mathbb{R} \cup \left(0, \tfrac{1}{2}\right) \cup \left(\tfrac{1}{2} + \mathbb{N}_0\right)\right) \times \prod_{j,\xi_j=1} (i\mathbb{R} \cup \mathbb{N}).$$

The *Bessel transform* $B_\xi^{\varepsilon,\varepsilon'} \varphi$ on $(\mathbb{R}^*)^d$, defined in (3.53), has product form
$$\left(B_\xi^{\varepsilon,\varepsilon'} \varphi\right)(t) = \prod_j \left(B_{\xi_j}^{\varepsilon_j, \varepsilon_j'} \varphi_j\right)(t_j),$$

with the local Bessel transform $B_{\xi_j}^{\varepsilon_j, \varepsilon_j'}$ in (3.55)–(3.58). Lemma 3.13 shows that the local Bessel transforms satisfy the estimates
$$(B_j \varphi_j)(t) \ll \min\left(|t|^{2\tau}, 1\right).$$

In the *Kloosterman sum*

$$S_\chi(r', r; c) = \sum_{a \bmod (c)}^* \chi(a) e^{2\pi i \mathrm{Tr}_{F/\mathbb{Q}}((ra+r'\tilde{a})/c)},$$

we let a run through representatives of $(O/I)^*$, and choose $\tilde{a} \in O$ such that $\tilde{a}a \equiv 1 \bmod (c)$. The Weil type estimate (2.47) implies that the following sum of Kloosterman sums converges absolutely:

$$K_\chi^{r,r'}\left(B_\xi^{\varepsilon,\varepsilon'}\varphi\right) = \sum_{c \in I \setminus \{0\}} \frac{S_\chi(r', r; c)}{|N(c)|} \left(B_\xi^{\varepsilon,\varepsilon'}\varphi\right)\left(\frac{4\pi|rr'|^{1/2}}{c}\right).$$

The argument of $B_\xi^{\varepsilon,\varepsilon'}\varphi$ is understood as $\left(\frac{4\pi|r_1 r_1'|^{1/2}}{c_1}, \ldots, \frac{4\pi|r_d r_d'|^{1/2}}{c_d}\right) \in (\mathbb{R}^*)^d$.

The first term on the right hand side of (3.84), called the *delta term*, vanishes unless there exists $\zeta \in O^*$ such that $r = \zeta^2 r'$ (hence $\varepsilon_j = \varepsilon'_j$ for $1 \le j \le d$). In this case, the delta term is equal to:

$$\frac{2\sqrt{|D_F|}}{(2\pi)^d} \chi(\zeta)^{-1} \prod_j (\mathrm{sign}\, \zeta_j)^{\xi_j}$$

$$\cdot \prod_{j, \xi_j = 0} \left(i \int_{\mathrm{Re}\, v = 0} \varphi_j(v)\, v \tan \pi v\, dv \right.$$

$$\left. + \sum_{b \ge 2,\, b \equiv 0 \bmod 2} (b-1)\varphi_j\left(\tfrac{b-1}{2}\right) \right)$$

$$\cdot \prod_{j, \xi_j = 1} \left(-i \int_{\mathrm{Re}\, v = 0} \varphi_j(v)\, v \cot \pi v\, dv \right.$$

$$\left. + \sum_{b \ge 3,\, b \equiv 1 \bmod 2} (b-1)\varphi_j\left(\tfrac{b-1}{2}\right) \right).$$

The left hand side of (3.84) depends on spectral data. It is described by a measure on Y_ξ depending on Fourier coefficients of automorphic forms for $\Gamma_0(I)$ with character $\begin{pmatrix} a & b \\ c & d \end{pmatrix} \mapsto \chi(d)$ of Γ, and central character given by $z_j \mapsto (-1)^{\xi_j}$, where $z_j \in \mathrm{SL}_2(\mathbb{R})^d$ is equal to $\begin{pmatrix} -1 & 0 \\ 0 & -1 \end{pmatrix}$ at place j, and equal to $\begin{pmatrix} 1 & 0 \\ 0 & 1 \end{pmatrix}$ at all other real places.

The definition of $d\sigma_{\chi,\xi}^{r,r'}$ in (3.43) consists of two terms. The first one gives the following contribution:

$$\sum_\varpi \overline{c^r(\varpi)}\, c^{r'}(\varpi) \varphi(v_\varpi).$$

The variable ϖ runs over a complete orthogonal system of irreducible cuspidal subspaces of $L_\xi^2\left(\Gamma_0(I) \backslash \mathrm{SL}_2(\mathbb{R})^d, \chi\right)$. We have chosen a spectral parameter $v_\varpi = (v_1, \ldots, v_d)$ for each ϖ such that $\frac{1}{4} - v_j^2$ is the eigenvalue in ϖ of the Casimir operator of the j-th factor of $\mathrm{SL}_2(\mathbb{R})^d$. We choose v_ϖ such that $\mathrm{Re}\, v_j \ge 0$. By $c^r(\varpi)$ we denote the Fourier coefficient of order r of the system $\{\psi_{\varpi,q}\}$ of automorphic forms in ϖ chosen in §2.3.4. With the normalization in (2.27), these Fourier coefficients do not depend on the choice of automorphic forms in ϖ.

The other term on the spectral side is due to the continuous spectrum. It has the form

$$\sum_\kappa c_\kappa \sum_{\mu \in \Lambda_{\kappa,\chi}} \int_{-\infty}^\infty \overline{D_\xi^r(\kappa, \chi; iy, i\mu)}\, D_\xi^{r'}(\kappa, \chi; iy, i\mu) \varphi(iy + i\mu)\, dy.$$

The variable κ runs over representatives of the finitely many $\Gamma_0(I)$-equivalence classes of cusps that are singular for the character $\begin{pmatrix} a & b \\ c & d \end{pmatrix} \mapsto \chi(d)$, i.e., χ restricted to $\Gamma \cap g_\kappa N g_\kappa^{-1}$ is trivial, where $g_\kappa \in G$ satisfies $\kappa = g_\kappa \infty$. The positive constants c_κ depend on the cusp, not on the character χ. At such a cusp κ there are Eisenstein series depending on quasi-characters of $(\mathbb{R}^*)^d / O^*$. These quasi-characters are specified by ν and μ, with $\nu \in \mathbb{C}$ and μ in a $(d-1)$-dimensional lattice $\Lambda_{\kappa,\chi}$. This determines $(\nu + i\mu_1, \ldots, \nu + i\mu_d) \in \mathbb{C}^d$. The $D_\xi^r(\kappa,\chi;\nu,i\mu)$ are Fourier coefficients of Eisenstein series, normalized in the same way as the $c^r(\varpi)$. See §2.2.3 and (2.31).

In Appendix B.1, we compare the sum formula (3.84) with Theorem 2.7.1 in [7], where the case of even weights and trivial character χ was treated.

Remark. The Bessel transformation is essential in the sum formula. It gives the relation between the test function on the spectral side to the test function in the sum of Kloosterman sums. Cogdell and Piatetski-Shapiro, [11], arrive directly at the kernel of the Bessel transformation.

We have used auxiliary test functions on G transforming on the left according to a character of N. The spectral theory of such functions is described by the Whittaker transformation in §3.2.1. So the Whittaker transformation is essential for our proof. Although interesting in its own right, it is not essential in many applications of the sum formula.

3.6.1. *Rational case.* In the case $d = 1$, the field F is $\mathbb{Q}$. We have $I = N\mathbb{Z}$ with $N \in \mathbb{N}$, and Γ is the Hecke congruence subgroup $\Gamma_0(N)$ of $SL_2(\mathbb{Z})$. The character χ of $(\mathbb{Z}/N)^*$ determines $\xi \in \{0,1\}$ by $\chi(-1) = (-1)^\xi$.

The g_κ in §2.1.4 can be chosen in $SL_2(\mathbb{Z})$. This gives each cusp κ a well determined width $w_\kappa \in \mathbb{N}$, determined by the fact that $g_\kappa n(w_\kappa) g_\kappa^{-1}$ generates $\Gamma \cap g_\kappa N g_\kappa^{-1}$. Another sensible choice is $\tilde{g}_\kappa = g_\kappa a(w_\kappa)$, such that $\tilde{g}_\kappa n(1) \tilde{g}_\kappa^{-1}$ generates $\Gamma \cap g_\kappa N g_\kappa^{-1}$. Note that $\tilde{g}_\kappa \in SL_2(\mathbb{R})$ need not be in $SL_2(\mathbb{Q})$.

For $d = 1$, the lattices $\Lambda_{\kappa,\chi}$ in §2.2.3 are zero, and the Eisenstein series depend only on $\nu \in \mathbb{C}$. The definition in (2.14) depends on the choice of g_κ. Let $\tilde{E}_q(\kappa,\chi;\nu)$ denote the Eisenstein series corresponding to the use of $\tilde{g}_\kappa$. Then

$$\tilde{E}_q(\kappa,\chi;\nu) = w_\kappa^{-\frac{1}{2}-\nu} E_q(\kappa,\chi;\nu).$$

If we work with the Eisenstein series $\tilde{E}_q$, the relation corresponding to (2.20), with f of weight q, is

$$(3.85) \qquad \|f^{\text{cont}}\|^2_{\Gamma \backslash \mathfrak{H}} = \frac{1}{4\pi} \sum_{\kappa \in \mathcal{P}_\chi} \int_{-\infty}^\infty \left| \langle f, \tilde{E}_q(\kappa,\chi;iy) \rangle \right|^2 dy.$$

See, e.g., Theorem 7.3 in [21], for the case $q = \xi = 0$. In the inner product $\langle f, \tilde{E}_q(\kappa,\chi;iy) \rangle$, we integrate over a fundamental domain for $\Gamma \backslash \mathfrak{H}$. This gives twice the integral over $\Gamma \backslash G$. Thus, we obtain for the present case the constants in (2.20):

$$(3.86) \qquad c_\kappa = \frac{1}{2\pi w_\kappa} \qquad (\kappa \in \mathcal{P}_\chi).$$

Let us reformulate the sum formula in the style of Theorem 1 in [12], with Fourier coefficients of individual automorphic forms. Instead of a sum over automorphic representations ϖ, there is a sum over orthonormal systems of automorphic forms. This amounts to choosing a suitable element in the representation space of each ϖ.

If ϖ is a unitary principal series or a complementary series representation, we choose a unit vector in the subspace with weight $q = \xi$. After numbering these representations in our orthogonal system, we obtain an orthonormal system (f_l) of real analytic cusp forms on $\mathfrak{H}$ with the following properties:

$$
\begin{aligned}
&\text{i)} \quad f_l(\gamma z) = \chi(d) e^{i\xi \arg(cz+d)} f_l(z) \text{ for } \gamma = \begin{pmatrix} a & b \\ c & d \end{pmatrix} \in \Gamma_0(N), \\
&\text{ii)} \quad \left(-y^2 \partial_y^2 - y^2 \partial_x^2 + i\xi \partial_x\right) f_l = \lambda_l f_l, \text{ with } \lambda_l > \xi/4, \\
&\text{iii)} \quad f_l(z) = {\sum_{r\in\mathbb{Z}}}' \rho_l(r) e^{2\pi i r x} W_{\frac{1}{2}\xi \operatorname{sign}(r), v_j}(4\pi |r| y), \text{ where } \lambda_l = \frac{1}{4} - v_l^2.
\end{aligned}
\tag{3.87}
$$

The eigenvalue 0, if present, does not correspond to a cusp form. The scalar product is given by integration over $\Gamma\backslash\mathfrak{H}$. On p. 9, we have mentioned a discrepancy by a factor 2 between the volumes of $\Gamma\backslash\mathfrak{H}$ and $\Gamma\backslash G$.

We use the norm $n(q, v)$ in (2.26) to relate the function f_l and the corresponding ϖ:

$$
v_l = v_\varpi, \quad f_l(z) = \frac{1}{\sqrt{2n(\xi, v_\varpi)}} \psi_{\varpi,\xi}(n(x) a(y)).
\tag{3.88}
$$

The Fourier coefficients satisfy

$$
\rho_l(r) = \frac{(-1)^\xi}{2\sqrt{|r| n(\xi, v_\varpi)} \, \Gamma\left(\frac{1}{2} + v_\varpi + \frac{\xi}{2} \operatorname{sign}(r)\right)} c^r(\varpi).
\tag{3.89}
$$

Here we have assumed that the freedom of a factor of absolute value one is used to relate the choice of the f_l to the choice of the system $(\psi_{\varpi,q})$; see (2.26). The sum formula does not depend on this choice.

If $\xi = 0$, the system (f_l) is an orthonormal basis of the cuspidal subspace. If $\xi = 1$, cusp forms with eigenvalue $\frac{1}{4}$ are not yet accounted for.

For ϖ a holomorphic discrete series representation, we choose a unit vector in the lowest weight space. If $v_\varpi = \frac{b-1}{2}$, $b \geq 1$, this lowest weight is equal to b. This leads to an orthonormal basis $(f_{b,l}^1)$ of the finite dimensional space of holomorphic cusp forms of weight b with character χ for each weight $b > 1$, $b \equiv \xi \bmod 2$.

$$
\begin{aligned}
&\text{i)} \quad f_{b,l}^1(\gamma z) = \chi(d)(cz+d)^b f_{b,l}^1(z) \text{ for } \gamma = \begin{pmatrix} * & * \\ c & d \end{pmatrix} \in \Gamma_0(N), \\
&\text{ii)} \quad f_{b,j}^1(z) = \sum_{r\in\mathbb{N}} \rho_{b,l}^1(r) e^{2\pi i r z}.
\end{aligned}
\tag{3.90}
$$

We use the Petersson scalar product:

$$
\int_{\Gamma\backslash\mathfrak{H}} \left|f_{b,l}^1(z)\right|^2 y^{b-2} \, dx\, dy = 1.
$$

The relation with the corresponding representation ϖ is as follows:

$$
f_{b,l}^1(z) = \frac{1}{\sqrt{2n(b, v_\varpi)}} y^{-b/2} \psi_{\varpi,b}(n(x) a(y)),
\tag{3.91}
$$

$$
\rho_{b,l}^1(r) = \frac{(-1)^b 2^{b-1} \pi^{b/2} r^{(b-1)/2}}{\sqrt{(b-1)!}} c^r(\varpi).
\tag{3.92}
$$

Similarly, the ϖ in the antiholomorphic discrete (or mock discrete) series with $v_\varpi = \frac{b-1}{2}$ correspond to an orthonormal basis $(f_{b,l}^{-1})$ of the space of antiholomorphic cusp forms

3. DERIVATION OF THE SPECTRAL SUM FORMULA

of weight b and character χ.

(3.93)   i) $f_{b,l}^{-1}(\gamma z) = \chi(d)(c\bar{z}+d)^b f_{b,l}^{-1}(z)$ for $\gamma = \begin{pmatrix} * & * \\ c & d \end{pmatrix} \in \Gamma_0(N)$,

(3.94)   ii) $f_{b,l}^{-1}(z) = \sum_{r \leq -1} \rho_{b,l}^{-1}(r) e^{2\pi i r \bar{z}}$.

Here as well, we use the Petersson scalar product.

$$(3.95) \qquad f_{b,l}^{-1}(z) = \frac{1}{\sqrt{2n(-b, v_\varpi)}} y^{-b/2} \psi_{\varpi, -b}(n(x) a(y)),$$

$$(3.96) \qquad \rho_{b,l}^{-1}(r) = \frac{(-1)^b 2^{b-1} \pi^{b/2} |r|^{(b-1)/2}}{\sqrt{(b-1)!}} c^r(\varpi).$$

Complex conjugation gives a bijection to holomorphic cusp forms of weight b and character χ^{-1}.

We use the Eisenstein series in weight $q = \xi$, which correspond to real-analytic automorphic forms $z \mapsto E_q(\kappa, \chi; v; n(x) a(y))$ on $\mathfrak{H}$ with the same transformation behavior as the f_l. The term of order $r \in \mathbb{Z} \setminus \{0\}$ in the Fourier expansion has the form $\rho(\kappa, r; v) e^{2\pi i r x}$ $W_{\frac{1}{2}\xi \operatorname{sign}(r), v}(4\pi |r| y)$, where

$$(3.97) \qquad \rho(\kappa, r; v) = \frac{(-1)^\xi}{\sqrt{2|r|} \Gamma\left(\frac{1}{2} + v + \frac{\xi}{2} \operatorname{sign}(r)\right)} D_\xi^r(\kappa, \chi; v).$$

In terms of these systems of automorphic forms on $\Gamma_0(N) \backslash \mathfrak{H}$, we have the following reformulation of Theorem 3.21 in the case $d = 1$:

PROPOSITION 3.22. *Let $r, r' \in \mathbb{Z} \setminus \{0\}$, and put $\varepsilon = \operatorname{sign} r$, $\varepsilon' = \operatorname{sign} r'$. Let χ be a character of $(\mathbb{Z}/N)^*$, and fix $\xi \in \{0, 1\}$ by $(-1)^\xi = \chi(-1)$. Let $\frac{1}{4} < \tau < \frac{1}{2}$.*

For each φ on $\{v \in \mathbb{C} : |\operatorname{Re} v| \leq \tau\} \cup \left(\frac{\xi-1}{2} + \mathbb{N}\right)$ that satisfies

    a) φ *is holomorphic on* $|\operatorname{Re} v| \leq \tau$,
    b) $\varphi(-v) = (\varepsilon \varepsilon')^\xi \varphi(n)$ *on* $|\operatorname{Re} v| \leq \tau$,
    c) $\varphi(v) \ll (1 + |v|)^{-a}$ *on its domain, for some $a > 2$,*

define the Bessel transform $B\varphi$ on $\mathbb{R}^*$ by (3.55)–(3.58).

The following equation holds, with all sums and integrals converging absolutely:

$$4\sqrt{|rr'|} \sum_l \varphi(v_l) \Gamma\left(\tfrac{1}{2} - v_l + \tfrac{\varepsilon\xi}{2}\right) \Gamma\left(\tfrac{1}{2} + v_l + \tfrac{\varepsilon'\xi}{2}\right) \overline{\rho_l(r)} \rho_l(r')$$

$$+ (\text{if } \varepsilon = \varepsilon') \sum_{b \geq 1, b \equiv \xi \bmod 2} \varphi\left(\tfrac{b-1}{2}\right) 2^{2-2b} \pi^{-b} |rr'|^{(1-b)/2} (b-1)!$$

$$\cdot \sum_l \overline{\rho_{b,l}^\varepsilon(r)} \rho_{b,l}^\varepsilon(r')$$

$$+ \frac{\sqrt{|rr'|}}{\pi} \sum_{\kappa \in \mathcal{P}_\chi} \frac{1}{w_\kappa} \int_{-\infty}^\infty \varphi(iy) \Gamma\left(\tfrac{1}{2} - iy + \tfrac{\varepsilon\xi}{2}\right) \Gamma\left(\tfrac{1}{2} + iy + \tfrac{\varepsilon'\xi}{2}\right)$$

$$\cdot \overline{\rho(\kappa, r; iy)} \rho(\kappa, r'; iy) \, dy$$

$$= \frac{\delta_{r,r'}}{\pi} \left(i \int_{\operatorname{Re} \nu = 0} \varphi(\nu) \nu \tan\left(\pi\left(\nu + \tfrac{\xi}{2}\right)\right) d\nu \right.$$

$$\left. + \sum_{b>1,\, b \equiv \xi \bmod 2} (b-1)\varphi\left(\tfrac{b-1}{2}\right) \right)$$

$$+ 2 \sum_{c \in \mathcal{NN}} (\mathrm{B}\varphi)\left(\tfrac{4\pi\sqrt{rr'}}{c}\right) \frac{1}{c} \sum_{a \bmod c}^{*} \chi(a) e^{2\pi i ((ra + r'\bar{a})/c)}.$$

See (3.87), (3.90), (3.94) and (3.97) for the definition of the various Fourier coefficients ρ.

In the Kloosterman term, we have taken together the (equal) contributions of c and $-c$.

In Appendix B.2, the sum formula in Proposition 3.22 is compared to that in [4].

3.6.2. *Full modular group.* A further specialization to $\xi = 0$, $N = 1$, gives the example in §1. Then $\Gamma = \Gamma_0(1)$ is equal to the modular group $\mathrm{SL}_2(\mathbb{Z})$. We take $r = n$ and $r' = m$ positive.

The f_l correspond to the u_j, and the $\rho_l(r)$ to $\gamma_j(r)$. We can arrange the f_l such that the Fourier coefficients have real values. There is only one cuspidal orbit, and the Fourier coefficients $c_{0,r}$ are known (see the explicit expression in (1.4)), and satisfy $\overline{c_{0,r}(\nu)} = c_{0,0}(-\nu)c_{0,r}(\nu)$ for $\operatorname{Re} \nu = 0$.

We take the test function $\varphi(\nu) = f(\nu)\frac{\cos \pi \nu}{\pi}$, with f as in (1.5), on the strip $|\operatorname{Re} \nu| \leq \tau$. The factor $\cos \pi \nu$ masks the contribution of the discrete series. With the integral representation (3.64):

$$\mathrm{B}_0^{1,1} \varphi(t) = \int_{\operatorname{Re} \nu = 0} f(\nu) 2\nu J_{2\nu}(|t|) \frac{d\nu}{2\pi i} = \tilde{f}(|t|).$$

Thus, we can check that Proposition 3.22 yields $4\sqrt{rr'}$ times the sum formula in (1.6).

4. Density results for cuspidal representations

As an application of the sum formula (3.84), we will obtain density results for automorphic representations, in the context of this paper. We shall follow the approach in [9].

We take in the situation of the sum formula $r = r'$. We partition the set of archimedean places of F:

$$(4.1) \qquad \{1, \ldots, d\} = E \sqcup Q_+ \sqcup Q_-,$$

where $Q_+ \cup Q_- \neq \emptyset$. We consider cuspidal automorphic representations ϖ in $L^2\left(\Gamma_0(I)\backslash G, \chi\right)$ that are restricted at the places in E, and are allowed to have a large spectral parameter at the other places, with the condition that it has discrete series type for $j \in Q_-$ and principal series or complementary series type for $j \in Q_+$:

$$(4.2) \qquad \mathcal{R} = \left\{ \varpi \text{ cuspidal} : \begin{cases} \lambda_{\varpi,j} \in [a_j, b_j] & \text{for } j \in E \\ \lambda_{\varpi,j} > 0 & \text{if } j \in Q_+ \\ \lambda_{\varpi,j} \leq 0 & \text{if } j \in Q_- \end{cases} \right\}.$$

The intervals $[a_j, b_j]$ can be anywhere in $\mathbb{R}$; for technical reasons we impose the condition that the end points a_j and b_j are not equal to discrete series or mock discrete series eigenvalues, i.e., not of the form $\frac{b}{2}\left(1 - \frac{b}{2}\right)$ with $b \geq 1$, $b \equiv \xi_j \bmod 2$. For the $\varpi \in \mathcal{R}$, we require that $\|\lambda_{\varpi,Q}\|_1 = \sum_{j \in Q_+ \cap Q_-} |\lambda_{\varpi,j}| \leq X$, with X large. So, we consider the weighted sum

(4.3) $$C^r(X) = \sum_{\varpi \in \mathcal{R}, \|\lambda_{\varpi,Q}\|_1 \leq X} |c^r(\varpi)|^2,$$

and compare it with the Plancherel measure. In (3.49) this measure is given in terms of the spectral parameter v. In the coordinate $\lambda = \frac{1}{4} - v^2$ it has the following description:

(4.4) $$\int_{Y_0} f(\lambda) \, d\widetilde{\mathrm{Pl}}_0(\lambda) = \int_{1/4} f(\lambda) \tanh\left(\pi \sqrt{\frac{1}{4} - \lambda}\right) d\lambda$$
$$+ \sum_{b \geq 2, \, b \equiv 0 \, (2)} (b-1) f\left(\frac{b}{2}\left(1 - \frac{b}{2}\right)\right),$$

$$\int_{Y_1} f(\lambda) \, d\widetilde{\mathrm{Pl}}_1(\lambda) = \int_{1/4} f(\lambda) \coth\left(\pi \sqrt{\frac{1}{4} - \lambda}\right) d\lambda$$
$$+ \sum_{b \geq 3, \, b \equiv 1 \, (2)} (b-1) f\left(\frac{b}{2}\left(1 - \frac{b}{2}\right)\right).$$

The main goal of this section is to prove:

THEOREM 4.1. *With these assumptions and notations:*

(4.5) $$\lim_{X \to \infty} C^r(X) X^{|E|-d} = \frac{2\sqrt{|D_F|}}{(d-|E|)! \, (2\pi)^d} \prod_{j \in E} \widetilde{\mathrm{Pl}}_{\xi_j}\left([a_j, b_j]\right).$$

We discuss some consequences of the theorem in special cases. Most of these consequences show the existence of infinitely many automorphic representations under restriction on the eigenvalue parameters at some places.

Weyl law weighted by Fourier coefficients of automorphic representations. Take $E = \emptyset$ and combine all possibilities for Q_+ and Q_-. Then the factor $(2\pi)^d$ in the denominator in (4.5) is replaced by π^d.

(4.6) $$\sum_{\varpi, \|\lambda_{\varpi,Q}\|_1 \leq X} |c^r(\varpi)|^2 \sim X^d \frac{2\sqrt{|D_F|}}{d! \, \pi^d} \quad (X \to \infty).$$

Infinitely many unitary principal series representations. We pick one place k, and restrict, at all other places, $\lambda_{\varpi,j}$ to an interval $[a_j, b_j] \subset \left(\frac{1}{4}, \infty\right)$.

(4.7) $$\sum_{\substack{\varpi, \, 0 \leq \lambda_{\varpi,k} \leq X \\ a_j \leq \lambda_{\varpi,j} \leq b_j \text{ for } j \neq k}} |c^r(\varpi)|^2$$
$$\sim X \frac{2\sqrt{|D_F|}}{(2\pi)^d} \prod_{j \neq k} \widetilde{\mathrm{Pl}}\left([a_j, b_j]\right) \quad (X \to \infty).$$

Low density of exceptional eigencoordinates. We look at representations that are exceptional at one place k. Theorem 4.1 implies:

(4.8) $$\sum_{\substack{\varpi, \, 0 \leq \lambda_{\varpi,k} \leq \frac{1}{4} \\ \sum_{j \neq k} |\lambda_{\varpi,j}| \leq X}} |c^r(\varpi)|^2 = o\left(X^{d-1}\right) \quad (X \to \infty).$$

We can even restrict the other $\lambda_{\varpi,j}$ further, letting only one eigencoordinate go to infinity. For $k \neq l$:

(4.9) $$\sum_{\substack{\varpi, 0 \leq \lambda_{\varpi,k} \leq \frac{1}{4}, |\lambda_{\varpi,l}| \leq X \\ \lambda_{\varpi,j} \in [a_j, b_j] \text{ for } j \neq k, l}} |c^r(\varpi)|^2 = o(X) \quad (X \to \infty).$$

Infinitely many representations of discrete series type. We pick a place k, and prescribe $\lambda_{\varpi,j} = \lambda_j = \frac{b_j}{2}\left(1 - \frac{b_j}{2}\right)$ with $b_j > 1$, $b_j \equiv \xi_j(2)$ for all places j except k, and let ϖ_k range freely through discrete series representations. Then we obtain:

(4.10) $$\sum_{\substack{\varpi, \lambda_{\varpi,j} = \lambda_j \text{ for } j \neq k \\ -X \leq \lambda_{\varpi,k} \leq 0}} |c^r(\varpi)|^2 \sim X \frac{\sqrt{|D_F|}}{\pi^d} \prod_{j \neq l} \sqrt{\frac{1}{4} - \lambda_j} \quad (X \to \infty).$$

Remark. These density results for discrete series type representation do not include the mock discrete series ($\lambda = \frac{1}{4}$, $\xi = 1$). The Plancherel measure $\coth\left(\pi\sqrt{\lambda - \frac{1}{4}}\right)$ in (4.4) connects the mock discrete series to the principal series.

4.1. Proof of Theorem 4.1. Much of the proof in §3-6 of [9] goes through. We discuss the main idea, skipping details when the argument in [9] goes through almost unchanged.

4.1.1. *Sum formula.* We choose the test function $\varphi = \bigotimes_j \varphi_j$ as follows:

(4.11)
$$j \in Q_+: \quad \varphi_j(\nu) = \begin{cases} e^{-s(\frac{1}{4} - \nu^2)} & \text{if } |\operatorname{Re}\nu| \leq \tau, \\ 0 & \text{otherwise;} \end{cases}$$

$$j \in Q_-: \quad \varphi_j(\nu) = \begin{cases} 0 & \text{if } |\operatorname{Re}\nu| \leq \tau, \\ e^{-s(\nu^2 - \frac{1}{4})} & \text{otherwise;} \end{cases}$$

$$j \in E: \quad \varphi_j \in T^1_{\xi_j}(\tau, a) \text{ arbitrary.}$$

As $s \downarrow 0$, the function $\varphi_j(\nu)$, $j \in Q_+$, approaches 1 on the spectral parameters of the unitary principal series and of the complementary series. Similarly, for $j \in Q_-$, the test function approaches 1 on the spectral parameters of the discrete series.

We insert φ into (3.84), and consider the behavior of the various terms as $s \downarrow 0$.

In [9] we worked with $\tau > \frac{1}{2}$. Here we use a narrow strip ($\frac{1}{4} < \tau < \frac{1}{2}$). In this way, we need not handle the discrete series eigenvalue $0 = \frac{2}{2}\left(1 - \frac{2}{2}\right)$ together with the continuous series eigenvalues.

Delta term. With (3.48), we find that the delta term is equal to

(4.12) $$\frac{2\sqrt{|D_F|}}{(2\pi)^d} \prod_j \int_{Y_{\xi_j}} \varphi_j \, d\mathrm{Pl}_{\xi_j}.$$

For $j \in Q_+$, using that $\tanh \pi t$ and $\coth \pi t$ behave like $1 + O\left(e^{-2\pi t}\right)$ as $t \to \infty$, we find

(4.13) $$\int_{Y_{\xi_j}} \varphi_j \, d\mathrm{Pl}_{\xi_j} = \frac{1}{s} + O(1) \quad (s \downarrow 0).$$

In §4.1 of [9], there is an additional term $\frac{1}{2}$, which is absorbed into $O(1)$. That is due to the condition $\tau > \frac{1}{2}$ there.

For $j \in Q_-$, we find $\frac{1}{s} + O(s^{-1/2})$ for the local Plancherel measure, by the method of (36) in [9]. For $\xi_j = 1$, the sum $e^{s/4} \sum_{m\geq 1} 2me^{-sm^2}$ arises. Approximation by an integral works again. The maximum of $x \mapsto 2xe^{-sx^2}$ is at $x = \frac{1}{2\sqrt{s}}$. The error in the approximation is $O(s^{-1/2})$. The integral itself is $\frac{1}{s}$.

Taking the product gives for the delta term:

$$(4.14) \qquad \frac{2\sqrt{|D_F|}}{(2\pi)^d} s^{|E|-d} \prod_{j\in E} \int_{Y_{\xi_j}} \varphi_j \, dPl_{\xi_j} \left(1 + o_E\left(s^{-1/2}\right)\right).$$

The subscript E in o_E, O_E and $\ll_E$ denotes dependence not only on the set E, but also on the choice of the test functions φ_j with $j \in E$.

Bessel transform. Lemma 3.12 gives

$$(4.15) \qquad \left(B_{\xi_j}^{\varepsilon_j,\varepsilon_j}\varphi_j\right)(t) \ll_{\varphi_j} \min\left(|t|^{2\tau}, 1\right),$$

which is the best we can do for $j \in E$.

At the places in $Q_+ \cup Q_-$, one can do better. Let first $j \in Q_+$. The formula giving $\varphi_j(\nu)$ for $|\operatorname{Re} \nu| \leq \tau$ extends holomorphically. We use (3.64) to obtain

$$(4.16) \qquad \left(B_{\xi_j}^{\varepsilon_j,\varepsilon_j}\varphi_j\right)(t) = 2\pi \left(i\varepsilon_j \operatorname{sign} t_j\right)^{\xi_j} \int_{\operatorname{Re}\nu=\gamma} e^{-s(\frac{1}{2}-\nu^2)} \frac{\nu J_{2\nu}(|t|)}{\cos\pi(\nu-\xi_j/2)} \frac{d\nu}{2\pi i}$$

for $\gamma \in \left(\tau, \frac{1}{2}\right)$. The estimate

$$(4.17) \qquad J_{2\nu}(y) \ll y^{2\gamma} e^{\pi|\operatorname{Im}\nu|}(1+|\operatorname{Im}\nu|)^{\frac{1}{2}-\tau-\gamma}$$

for $y > 0$ on p. 700 of [9] goes through for $\operatorname{Re}\nu = \gamma$. This gives

$$(4.18) \qquad \left(B_{\xi_j}^{\varepsilon_j,\varepsilon_j}\right)(t) \ll \int_{-\infty}^{\infty} e^{-s(u^2-\gamma^2)}|t|^{2\gamma}(1+|u|)^{\frac{1}{2}-\tau-\gamma}|u|\,du$$

$$\ll |t|^{2\tau} s^{\frac{\gamma+\tau}{2}-\frac{5}{4}}.$$

For large t, we use $J_{2\nu}(y) \ll e^{\pi|\operatorname{Im}\nu|}$ for $\operatorname{Re}\nu \geq \frac{1}{4}$, $y > 0$, see (3.66), to obtain for $j \in Q_+$:

$$(4.19) \qquad \left(B_{\xi_j}^{\varepsilon_j,\varepsilon_j}\varphi_j\right)(t) \ll \min\left(|t|^{2\tau} s^{\frac{\gamma+\tau}{2}-\frac{5}{4}}, \frac{1}{s}\right).$$

Let $j \in Q_-$. In the way shown at the bottom of p. 700 of [9], we have $J_{b-1}(y) \ll y^{2\tau}b^{-2\tau}$. Hence

$$(4.20) \qquad \left(B_{\xi_j}^{\varepsilon_j,\varepsilon_j}\varphi_j\right)(t) \ll |t|^{2\tau} \sum_{m\geq 2} \left(\frac{\xi_j}{2}+2m\right)^{1-2\tau} e^{-s\left(\frac{\xi_j}{2}+m\right)\left(\frac{\xi_j}{2}-1+m\right)}$$

$$\ll |t|^{2\tau} s^{\tau-1}.$$

From $|J_{b-1}(y)| \leq 1$, it follows that $\left(B_{\xi_j}^{\varepsilon_j,\varepsilon_j}\varphi_j\right)(t) \ll s^{-1}$.

This implies that there is $\zeta \in (-1,0)$ such that

$$(4.21) \qquad \left(B_{\xi}^{\varepsilon,\varepsilon}\varphi\right)(t) \ll_E \prod_j \left(s^{\alpha_j}|t_j|^{2\tau}, s^{-\beta_j}\right),$$

$$\alpha_j = \begin{cases} 0 & (j \in E), \\ \zeta & (j \in Q_+ \cap Q_-); \end{cases} \qquad \beta_j = \begin{cases} 0 & (j \in E), \\ 1 & (j \in Q_+ \cup Q_-). \end{cases}$$

Kloosterman term. The Weil estimate (2.47) implies that for each $\delta > 0$:

$$(4.22) \quad K_\chi^{r,r}\left(B_\xi^{\varepsilon,\varepsilon}\varphi\right) \ll_{F,r,\delta,E} \sum_{c \in I}{}' N(J_c)^{-\frac{1}{2}+\delta} \prod_j \min\left(s^{\alpha_j}\left(\frac{4\pi|r_j|}{|c_j|}\right)^{2\tau}, s^{-\beta_j}\right)$$

$$= \sum_{(c) \text{ princ. id.}} N(J_c)^{-\frac{1}{2}+\delta} \sum_{\varepsilon \in O^*} \min\left(\frac{s^{\alpha_j}(4\pi|r_j|)^{2\tau}|^{2\tau}}{|\varepsilon_j c_j|^{2\tau}}, s^{-\beta_j}\right).$$

We apply Lemma 2.2 to the inner sum with $\alpha = 2\tau$, $\beta = 0$, $y = \frac{1}{c}$ and

$$(4.23) \quad p_j = s^{\alpha_j}\left(4\pi|r_j|\right)^{2\tau}, \quad P = (4\pi)^{2\tau d}|N(r)|^{2\tau}s^{\zeta(d-|E|)},$$
$$q_j = s^{\beta_j}, \quad Q = s^{s-|E|},$$

to find the estimate

$$\ll_\tau \left(1 + \log|N(c)| + \left|\log\left((4\pi)^{2\tau d}|N(r)|^{2\tau}s^{(1-\zeta)(|E|-d)}\right)\right|\right)^{d-1}$$
$$\cdot \min\left(\frac{(4\pi)^{2\tau d}|N(r)|^{2\tau}s^{\zeta(d-|E|)}}{|N(c)|^{2\tau}}, s^{d-|E|}\right)$$
$$\ll_{d,E,r,\delta} |N(c)|^\delta s^{-\delta+\zeta(d-|E|)}|N(c)|^{-2\tau+\delta}.$$

(For our purpose, it suffices to use only the first argument in the minimum.)

We find the following estimate of the Kloosterman term:

$$(4.24) \quad K_\chi^{r,r}\left(B_\xi^{\varepsilon,\varepsilon}\varphi\right) \ll_{r,\tau,d,\delta,E} \sum_{(c)} N(J_c)^{-\frac{1}{2}+\delta} s^{-\delta+\zeta(d-|E|)}|N(c)|^{-2\tau+\delta}.$$

This converges, if we take $\delta > 0$ sufficiently small. We see this from a product expansion in prime ideals:

$$(4.25) \quad K_\chi^{r,r}\left(B_\xi^{\varepsilon,\varepsilon}\varphi\right) \ll_{r,\tau,d,\delta,E} s^{-\delta+\zeta(d-|E|)} \prod_{P|I}\left(1 - N(P)^{-2\tau+\delta}\right)^{-1}$$
$$\cdot \prod_{P\nmid I}\left(1 - N(P)^{-\frac{1}{2}-2\tau+2\delta}\right)^{-1} = O_{E,r}\left(s^{-\delta+\zeta(d-|E|)}\right).$$

Since $\zeta \in (-1, 0)$ and δ can be taken sufficiently small, this is smaller than the delta term in (4.14).

Spectral term. The sum formula gives

$$(4.26) \quad \int_{Y_\xi} \varphi(\nu)\, d_{\chi,\xi}^{r,r}(\nu) = \frac{2\sqrt{|D_F|}}{(2\pi)^d} s^{|E|-d} \prod_{j \in E}\left(\int_{Y_{\xi_j}} \varphi_j d\mathrm{Pl}_{\xi_j}\right)(1 + o_{E,r}(1)),$$

where o_E denotes not only the dependence on the set E, but also on the choice of the test functions φ_j with $j \in E$.

Equation (3.43) shows that the spectral term is the sum of the following two terms:

$$(4.27) \quad \text{Cuspidal term:} \quad \sum_\varpi \varphi(\nu_\varpi)|c^r(\varpi)|^2,$$

$$\text{Eisenstein term:} \quad 2 \sum_{\kappa \in \mathcal{P}_\chi} c_\kappa \sum_{\mu \in \Lambda} \int_0^\infty \varphi(i y + i\mu)|D^r(\kappa, \chi; iy, i\mu)|^2\, dy.$$

The cuspidal term is similar to the quantity C^r we are after. The Eisenstein term is zero if $Q_- \neq \emptyset$.

The D^r in the Eisenstein term are Fourier coefficients of Eisenstein series, see §2.2.3 and (2.31). In the present arithmetic situation, there are good estimates for these Fourier coefficients. This is discussed in §4.2, where we show how the reasoning in §5.1 of [9] goes through for odd weights and non-trivial characters.

With Proposition 4.2 we find the following estimate for the Eisenstein term:

$$(4.28) \quad \ll_r \sum_\kappa \sum_{\mu \in \Lambda_{\kappa,\chi}} \int_0^\infty \prod_{j \in E} \varphi_j(iy + i\mu_j) \prod_{j \in Q_+} e^{-s((y+\mu_j)^2 + \frac{1}{4})}$$
$$\cdot \left((\log(2+y))^7 + (\text{if } \mu \neq 0) \left(\log \max |\mu_j|\right)^7 \right)^2 dy.$$

The $\Lambda_{\kappa,\chi}$ are lattices in the hyperplane $\sum_j x_j = 0$ in $\mathbb{R}^d$. We estimate these sums by an integral over this hyperplane. We use $\varphi_j(ix) \ll_E (1+|x|)^{-a}$, $a > 2$, for the $j \in E$. As in (77) in [9], we arrive at $O_{\varepsilon,E}\left(s^{-\frac{1+\varepsilon}{2}(d-|E|)}\right)$ as estimate of the Eisenstein term, for each $\varepsilon > 0$. This has smaller growth than the main term in (4.26).

In this way, we arrive at

$$(4.29) \quad \sum_{\substack{\varpi, \, \lambda_{\varpi,j} > 0 \text{ for } j \in Q_+ \\ \lambda_{\varpi,j} \leq 0 \text{ for } j \in Q_-}} |c^r(\varpi)|^2 \, e^{-s \sum_{j \in Q_+ \cup Q_-} |\lambda_{\varpi,j}|} \prod_{j \in E} \varphi_j(\nu_{\varpi,j})$$

$$\sim s^{|E|-d} \frac{2\sqrt{|D_F|}}{(2\pi)^d} \prod_{j \in E} \left(\int_{Y_{\xi_j}} \varphi_j \, d\text{Pl}_{\xi_j} \right) \quad \text{as } s \downarrow 0.$$

This proves the analogue of Theorem 3.1 in [9], after the introduction, for $j \in E$, of functions g_j for $j \in E$, defined by $\varphi_j(\nu) = g_j\left(\frac{1}{4} - \nu^2\right)$.

4.1.2. *Tauberian argument and approximation.* The reasoning in §3 of [9] carries over almost unchanged. First a Tauberian argument is used to obtain

$$(4.30) \quad \sum_{\substack{\varpi, \, \|\lambda_{\varpi,Q}\|_1 \leq X \\ \lambda_{\varpi,j} > 0 \text{ for } j \in Q_+ \\ \lambda_{\varpi,j} \leq 0 \text{ for } j \in Q_-}} |c^r(\varpi)|^2 \prod_{j \in E} g_j\left(\lambda_{\varpi,j}\right)$$

$$\sim X^{d-|E|} \frac{2\sqrt{|D_F|}}{(d-|E|)! \, (2\pi)^d} \prod_{j \in E} \left(\int_{Y_{\xi_j}} \varphi_j \, d\text{Pl}_{\xi_j} \right) \quad \text{as } X \to \infty.$$

(See Proposition 3.2 in [9].)

The final step, see the proof of Theorem 3.3 in [9], is the approximation of the characteristic function of the interval $[a_j, b_j]$ by g_j. This goes through in the present situation. The function b_j at the bottom of p. 690 should be taken in a slightly different way: take $\frac{1}{4} - \tau^2 \in \left(0, \frac{3}{16}\right)$ as the separation point. At the top of p. 691, b_j should be extended to a neighborhood of $\left[\frac{1}{4} - \tau^2, \infty\right)$.

4.2. Fourier coefficients of Eisenstein series. The contribution of the continuous spectrum to the sum formula is based on the normalized Fourier coefficients $D^r_\xi(\kappa, \chi)$ in (2.31). Sorensen, [45], gives an explicit description of these coefficients for $\Gamma = \text{SL}_2(O)$ and $\xi = 0$. For congruence subgroups an explicit description gets more complicated. In this subsection, we sketch how the discussion in §5 of [9] applies in the present situation, and yields estimates for $D^r_\xi(\kappa, \chi; \nu, i\mu)$ on the line $\text{Re}\,\nu = 0$. We have used such an estimate to obtain Theorem 4.1.

This discussion also shows that the constant functions are the sole square integrable automorphic forms orthogonal to the spaces of cusp forms in $L^2_\xi(\Gamma\backslash G,\chi)$.

The Eisenstein series $E_q(\kappa,\chi)$ for $\Gamma_0(I)$ with character χ are linear combinations of Eisenstein series for the principal congruence subgroup

$$\Gamma(I) = \left\{ \begin{pmatrix} a & b \\ c & d \end{pmatrix} \in SL_2(O), \begin{pmatrix} a & b \\ c & d \end{pmatrix} \equiv \begin{pmatrix} 1 & 0 \\ 0 & 1 \end{pmatrix} \mod I \right\}.$$

So growth estimates for Fourier coefficients of Eisenstein series can be transported from $\Gamma(I)$ to $\Gamma_0(I)$. We look at Eisenstein series for $\Gamma(I)$ of weight ξ, and take the factor in front of $d^r(\xi, \nu+i\mu)W^r_\xi(a(y))$ in the Fourier terms of order r. For $\Gamma(I)$ the Fourier term orders at ∞ run through the fractional ideal $I^{-1}O'$. For $\Gamma_1(I)$ and $\Gamma_0(I)$ we need only $r \in O' \setminus \{0\}$.

Section 5.1 of [9] discusses the Eisenstein series for $\Gamma(I)$ at a cusp $-\frac{\delta}{\gamma}$, $\gamma, \delta \in O$. The influence of a more general weight q (that has to be congruent to ξ mod $2\mathbb{Z}^d$) is felt in (47) in *loc. cit.*, where a factor

$$\prod_j e^{-iq_j \arg(c_j z_j + d_j)}$$

has to be inserted. This propagates to the next displayed equation, where

$$d^r_\infty(0, \nu + i\mu)W^{r,\nu+i\mu}_{\infty,0}(a(y))$$

has to be replaced by $d^r(q, \nu + i\mu)W^r_\xi(a(y))$. In view of (2.31), the influence of the weight is not felt by the Dirichlet series in front of these factors. This is the essential part of the normalized Fourier coefficient.

Then the reasoning goes on unchanged, and leads to the following structure for the Fourier coefficient of the $\Gamma(I)$-Eisenstein series at the cusp $\lambda = -\frac{\delta}{\gamma}$:

$$\sum_\tau (\text{constant}) \cdot \left(\text{finite sum of terms } \prod_j t_j^{\nu+i\mu_j} \text{ with } t_j > 0 \right)$$

$$\cdot \sum_{\chi_1} \overline{\chi_1(\tau)} \frac{1}{L(1 + 2\nu, \bar\lambda_\mu, \chi_1)},$$

where τ runs over the (finite) ray class group

$$\{\mathfrak{b} \text{ ideal of } O \text{ prime to } I\} / \{(b) : b \equiv 1 \mod I, b \text{ totally positive}\},$$

and where χ_1 runs over the characters of this group.

The L-function $L(s, \bar\lambda_\mu, \chi_1)$ is built with the character $\chi_1\bar\lambda_\mu$ of the ideals prime to I. The character λ_μ is an extension of $\vartheta \mapsto \prod_j \vartheta_j^{2i\mu_j}$.

This L-series has a convergent Euler product for Re $s > 1$. This shows that the Eisenstein series have no singularities for $0 < \nu < \frac{1}{2}$.

Proposition 5.2 of [9] gives the bound

(4.31) $\qquad \dfrac{1}{L(s,\bar\lambda_\mu,\chi)} \ll_{F,I} \begin{cases} \log^7(2 + |\text{Im } s|) + \log^7 \|\mu\| & \text{if } \mu \neq 0, \\ \log^7(1 + |\text{Im } s|) & \text{if } \mu = 0, \end{cases}$

for Re $s = 1$. We use $\|\mu\| := \max_j (|\mu_j| + 1)$.

The Fourier coefficients of Eisenstein series for $\Gamma_0(I)$ are linear combinations of Fourier coefficients of Eisenstein series for $\Gamma(I)$. The absolute value of the coefficients in this linear combination depend on the character χ of $\Gamma_0(I)$, and are bounded for Re $\nu = 0$. This leads to the following result, which we used to prove (4.29):

PROPOSITION 4.2. *For $\kappa \in \mathcal{P}_\chi$, $\operatorname{Re} \nu = 0$, $\mu \in \Lambda_{\kappa,\chi}$, $\varepsilon > 0$:*

$$D_\xi^r(\kappa,\chi;\nu,i\mu) \ll_{F,I,\varepsilon} |N(r)|^\varepsilon \begin{cases} \log^7(2+|\nu|) + \log^7 \|\mu\| & \text{if } \mu \neq 0, \\ \log^7(2+|\nu|) & \text{if } \mu = 0. \end{cases}$$

See (76) in [**9**].

The proof in §5.2 in [**9**] uses a method of Landau, and does not depend on the functional equation of the L-series $L(s, \bar{\lambda}_\mu, \chi_1)$. Y.Motohashi has pointed out to us another proof, based on the Lemmas α, β, γ in 3.9 of [**46**], which leads to an estimate with a lower exponent on the logarithms.

CHAPTER 2

Kloosterman sum formula

As we have seen in §4, Theorem 3.21 is useful to obtain information concerning spectral data: the ν_ϖ and the $c^r(\varpi)$. However, to investigate sums of Kloosterman sums $K_\chi^{r,r'}\left(B_\xi^{\varepsilon,\varepsilon'}\varphi\right)$, it is preferable to have $B_\xi^{\varepsilon,\varepsilon'}\varphi$ as the independent test function. Our aim is now the derivation of such a formula, namely (6.44) in Theorem 6.5, which is the other main result in this paper.

5. Bessel inversion

In view of (3.53), Bessel inversion can be done place by place, by the inversion of the transformations in (3.55)–(3.58). In the case $\varepsilon_j\varepsilon'_j = -1$, (3.57), (3.58), we have the Kontorovitch-Lebedev transform, see [**32**], p. 131. Kuznetsov discusses the Bessel transform in the case $\varepsilon_j\varepsilon'_j = 1$, (3.55), (3.56) (see p. 14 and Theorem 5 in [**28**], and p. 368, the theorem in the appendix of [**29**] and references therein). He gives the result in a generality that makes it suitable for the sum formula on the universal covering group of $SL_2(\mathbb{R})$, as discussed in [**41**] and [**4**].

We give the main line of the proof, as the Bessel inversion is essential for the study of sums of Kloosterman sums by means of the sum formula. The main idea of the proof works also for $PSL_2(\mathbb{C})$, see [**10**]. The flavor of the proof is a manipulation of power series. In the work of Motohashi, see especially [**38**], the Mellin transform of Bessel functions is used profitably to work with the Bessel transform in the sum formula.

We note that $SL_2(\mathbb{R})$ and $SL_2(\mathbb{C})$ are the sole Lie groups for which the Bessel inversion is known; see [**10**] and [**33**] for the case of $SL_2(\mathbb{C})$.

5.1. Bessel transform on $SL_2(\mathbb{R})$. We consider the transform $B = B_\xi^{\varepsilon,\varepsilon'}$ in equations (3.55)–(3.58), with $\varepsilon, \varepsilon' \in \{1, -1\}$, $\xi \in \{0, 1\}$. We omit the parameters $\varepsilon, \varepsilon'$ and ξ from the notation in this subsection.

The transform B connects functions on $i\mathbb{R} \cup \left(\frac{\xi-1}{2} + \mathbb{N}_0\right)$ that satisfy $\varphi(-\nu) = (\varepsilon\varepsilon')^\xi\varphi(\nu)$ for $\nu \in i\mathbb{R}$ to functions on $\mathbb{R}^*$ that obey the symmetry condition $f(-t) = (-1)^\xi f(t)$.

$$(5.1) \qquad B\varphi(t) = \int_{Y_\xi} \varphi(\nu) k_\xi^{\varepsilon,\varepsilon'}(\nu, t)\, d\mathrm{Pl}_\xi(\nu),$$

with $k_\xi^{\varepsilon,\varepsilon'}$ as in (3.35). The test functions in Definition 3.10 are integrable and square integrable for $d\mathrm{Pl}_\xi$. This operator B is $(-\varepsilon)^\xi$ times the operator $\left(k_\xi^{\varepsilon\varepsilon'}\right)^{\leftarrow}$ in [**4**], Proposition 14.2.8.

The transpose operator is:

$$(5.2) \qquad B^*f(\nu) := \int_{\mathbb{R}^*} f(t)\overline{k_\xi^{\varepsilon,\varepsilon'}(\bar\nu, t)}\, \frac{dt}{|t|}$$
$$= 2(-1)^\xi \int_0^\infty f(t) k_\xi^{\varepsilon,\varepsilon'}(\nu, t)\, \frac{dt}{t}.$$

Let now $f \in C_c^\infty(\mathbb{R}^*)_\xi$. (The subscript ξ signals the parity condition $f(-t) = (-1)^\xi f(t)$.)
The integral in (5.2) converges absolutely and defines $\mathrm{B}^* f(\nu)$ as a holomorphic function of $\nu \in \mathbb{C}$.

With the Bessel transform

$$(5.3) \qquad j^{\pm 1} f(\nu) := \int_0^\infty f(t) J_\nu^{\pm 1}(t) \, \frac{dt}{t},$$

we have

$$(5.4) \qquad \mathrm{B}^* f(\nu) = 2(i\varepsilon)^\xi \cos \pi \left(\nu + \tfrac{\xi}{2}\right) \frac{(-\varepsilon\varepsilon')^\xi j^{\varepsilon\varepsilon'} f(-2\nu) - j^{\varepsilon\varepsilon'} f(2\nu)}{\sin 2\pi\nu},$$

$$\mathrm{B}^* f\left(\tfrac{b-1}{2}\right) = \begin{cases} 2\varepsilon^\xi i^b j^1 f(b-1) & \text{if } \varepsilon\varepsilon' = 1, \\ 0 & \text{if } \varepsilon\varepsilon' = -1. \end{cases}$$

The former equality is useful for general values of ν; in the latter, we suppose that $b \equiv \xi \bmod 2$. The function $\mathrm{B}^* f$ here corresponds to $(-\varepsilon')^\xi b_\xi^{\varepsilon\varepsilon'} f$ in (14.2.9), [4].

The Bessel transform $j^{\pm 1}$ has the following expression in terms of Mellin transforms:

$$(5.5) \qquad j^{\pm 1} f(\nu) = \sum_{n=0}^\infty \frac{(\mp 1)^n}{2^{2\nu+2n} \, n! \, \Gamma(1 + 2\nu + n)} \mathcal{M} f(2n + 2\nu),$$

$$\mathcal{M} f(s) := \int_0^\infty f(t) t^s \, \frac{dt}{t}.$$

With standard estimates of Mellin transforms of smooth compactly supported functions, we have for ν with $\mathrm{Re}\,\nu \geq -\tau$ and for all $l \in \frac{1}{2}\mathbb{Z}$:

$$(5.6) \qquad j^{\varepsilon\varepsilon'} f(\nu) \ll_\varepsilon \sum_{n \geq 0} \frac{|\mathcal{M} f(2n + 2\nu)|}{4^{n+\mathrm{Re}\,\nu} |\Gamma(1 + 2\nu)|} \ll_{l,f} \frac{C_f^{|\mathrm{Re}\,\nu|}}{|\Gamma(1 + 2\nu)| \, (1 + |\mathrm{Im}\,\nu|)^l},$$

for each $l \geq 0$.

LEMMA 5.1. *If $f \in C_c^\infty(\mathbb{R}^*)_\xi$, $a > 2$, and $\tau \in \left(\tfrac{1}{4}, 1\right)$, then $\mathrm{B}^* f \in T_\xi^{\varepsilon\varepsilon'}(\tau, a)$ (see Definition 3.10).*

The question now is what $\mathrm{BB}^* f$ is. If it is a multiple of f then we have available a class of test functions in the Kloosterman term that suffices for many practical purposes.

5.2. Scalar products. We continue considering one place. So $\xi \in \{0, 1\}$. The scalar product

$$(5.7) \qquad \int_{Y_\xi} \mathrm{B}^* f_1(\nu) \, \overline{\mathrm{B}^* f_2(\nu)} \, d\mathrm{Pl}_\xi(\nu),$$

for $f_1, f_2 \in C_c^\infty(\mathbb{R}^*)_\xi$, is the sum of an integral I over $i\mathbb{R}$ and a sum D over $b \equiv \xi \bmod 2$, see (3.49). The estimate (5.6) shows that the following contribution, based on (3.49) and (5.4), converges absolutely:

$$(5.8) \qquad D = \begin{cases} 4 \sum_{a \geq 1, a \not\equiv \xi \bmod 2} a \, j^1 f_1(a) \, \overline{j^1 f_2(a)} & \text{if } \varepsilon\varepsilon' = 1, \\ 0 & \text{if } \varepsilon\varepsilon' = -1. \end{cases}$$

The integral can be written as follows:

$$I = -2(i\varepsilon)^\xi \int_{\mathrm{Re}\,\nu=0} \frac{\cos \pi \left(\nu + \tfrac{\xi}{2}\right)}{\sin 2\pi\nu} j^{\varepsilon\varepsilon'} f_1(2\nu) \, \overline{\mathrm{B}^* f_2(\nu)} \, d\mathrm{Pl}_\xi(\nu).$$

5. BESSEL INVERSION

The zero of $\sin 2\pi\nu$ is compensated by either $\cos\pi\left(\nu + \frac{\xi}{2}\right)$ or by a zero of the density of $d\mathrm{Pl}_\xi$. Breaking up $B^* f_2$ as well, we arrive at:

(5.9) $\qquad I = I_1 + I_2$,

$$I_1 = 8\pi \int_{\mathrm{Re}\,\nu=0} j^{\varepsilon\varepsilon'} f_1(2\nu) \, \overline{j^{\varepsilon\varepsilon'} f_2(2\nu)} \, \frac{\nu}{\sin 2\pi\nu} \frac{d\nu}{2\pi i},$$

$$I_2 = -8\pi(-\varepsilon\varepsilon')^\xi \int_{\mathrm{Re}\,\nu=0} j^{\varepsilon\varepsilon'} f_1(2\nu) \, \overline{j^{\varepsilon\varepsilon'} f_2(-2\nu)} \, \frac{\nu}{\sin 2\pi\nu} \frac{d\nu}{2\pi i}$$

Replacing the Bessel transforms in I_1 by the initial term of the expansion (5.5) gives:

(5.10) $\qquad I_1^0 = 4 \int_{\mathrm{Re}\,\nu=0} \mathcal{M}f_1(2\nu) \, \overline{\mathcal{M}f_2(2\nu)} \, \frac{d\nu}{2\pi i} = \int_{\mathbb{R}^*} f_1(t) \, \overline{f_2(t)} \, \frac{dt}{|t|}.$

We look at the remaining terms. As $X \to \infty$:

(5.11) $\qquad I_1 - I_1^0 = \int_0^\infty \int_0^\infty k_X(t_1, t_2) f_1(t_1) \overline{f_2(t_2)} \, \frac{dt_2}{t_2} \frac{dt_1}{t_1} + o(1),$

where

(5.12) $\qquad k_X(t_1, t_2) = \sum_{n,m \geq 0}' \int_{-iX}^{iX} \frac{(-\varepsilon\varepsilon')^{n+m}(t_1/2)^{2n+2\nu}(t_2/2)^{2m-2\nu}}{n!\,m!\,\Gamma(1+2\nu+n)\Gamma(1-2\nu+m)}$
$$\cdot \frac{8\pi\nu}{\sin 2\pi\nu} \frac{d\nu}{2\pi i}.$$

By $\sum'$ we mean that $(n, m) = (0, 0)$ is omitted from the sum. We take $X \in \frac{1}{4} + \mathbb{N}$.

We assume for the moment that $t_1 \leq t_2$. The integral from $-iX$ to iX along $i\mathbb{R}$ can be shifted into a half circle in the right half plane. Outside small neighborhoods of points in $\frac{1}{2}\mathbb{Z}$, we have

$$\frac{1}{\Gamma(1+2\nu+n)\Gamma(1-2\nu+m)} \ll \frac{1}{|(2\nu-1)\Gamma(1+2\nu)\Gamma(1-2\nu)|}$$
$$= \frac{|\sin 2\pi\nu|}{2\pi|\nu|\,|2\nu-1|}.$$

The integral of one term over the half circle satisfies the estimate

$$\ll \int_{-\pi/2}^{\pi/2} \left(\frac{t_1}{t_2}\right)^{2X\cos\vartheta} \frac{(t_1/2)^{2n}(t_2/2)^{2m} X\,d\vartheta}{n!\,m!\,(X-1)}$$
$$\ll \frac{(t_1/2)^{2n}(t_2/2)^{2m}}{n!\,m!} \min\left(1, \frac{1}{X\log(t_2/t_1)}\right).$$

To estimate the integral against $f_1(t_1)\overline{f_2(t_2)}$, we write $t_1 = pq$, $t_2 = p/q$, and take Q_0 and Q_1 such that the support of $f_1 \bar{f}_2$ is contained in $[Q_0, Q_1]^2$ in the (p, q)-plane. We obtain the following estimate of the contribution to $I_1 - I_1^0$ from the region $t_1 \leq t_2$:

$$\int_{p=0}^\infty \int_{q=1}^\infty |f_1(pq) f_2(p/q)| \, e^{p^2(q^2+q^{-2})/4} \min\left(1, \frac{1}{X\log q}\right) \frac{dq}{q} \frac{dp}{p}$$

$$\ll A \int_{p=Q_0}^{Q_1} \left(\int_{q=1}^{e^{1/X}} \frac{dq}{q} + \int_{q=e^{1/X}}^{Q_1} \frac{dq}{qX\log q} \right) \frac{dp}{p}$$

(5.13) $\qquad \ll A \log \frac{Q_1}{Q_0} \left(\frac{1}{X} + \frac{\log\log Q_1 + \log X}{X} \right) = o(1),$

where A depends on the size of $f_1 \overline{f_2}$.

The residues contribute the following to the term of order (n, m):

$$-2 \sum_{1 \leq a < 2X, a \leq m} \frac{(-\varepsilon\varepsilon')^{n+m}(-1)^a a (t_1/2)^{2n+a}(t_2/2)^{2m-a}}{n!\, m!\, (a+n)!\, (m-a)!}.$$

After adding over $(n, m) \neq (0, 0)$, we obtain

(5.14) $$-2 \sum_{1 \leq a < 2X} (\varepsilon\varepsilon')^a a J_a^{\varepsilon\varepsilon'}(t_1) J_a^{\varepsilon\varepsilon'}(t_2).$$

On the region $t_1 \geq t_2$, we proceed similarly, using a half circle in the region $\operatorname{Re} \nu \leq 0$. The contribution of the integral over the half circle is again $o(1)$. The residues contribute again:

(5.15) $$-2 \sum_{1 \leq a < 2X} (\varepsilon\varepsilon')^a a J_a^{\varepsilon\varepsilon'}(t_1) J_a^{\varepsilon\varepsilon'}(t_2).$$

These sums converge absolutely on each compact set in $(0, \infty)^2$. We conclude, after integration against $f_1(t)\overline{f_2(t)}$, that

(5.16) $$I_1 - I_1^0 = -2 \sum_{a=1}^{\infty} (\varepsilon\varepsilon')^a a j^{\varepsilon\varepsilon'} f_1(a) \overline{j^{\varepsilon\varepsilon'} f_2(a)}.$$

From (5.5) it is clear that $\overline{j^{\varepsilon\varepsilon'} f_2(w)} = j^{\varepsilon\varepsilon'} \overline{f_2}(\bar{w})$. We shift the line of integration in the expression for I_2 in (5.9) to the right. Estimate (5.6) shows that the integral tends to zero. We are left with a sum of residues:

(5.17) $$I_2 = 2(-\varepsilon\varepsilon')^\xi \sum_{a=1}^{\infty} (-1)^a a\, j^{\varepsilon\varepsilon'} f_1(a)\, j^{\varepsilon\varepsilon'} \overline{f_2}(a).$$

The conclusion is that $I_1 - I_1^0 + I_2$ is equal to $-D$, see (5.8). We have shown:

PROPOSITION 5.2. *For all $f \in C_c^\infty (\mathbb{R}^*)_\xi$, the function $\mathrm{B}^* f$ is square integrable for $d\mathrm{Pl}_\xi$, and*

(5.18) $$\int_{Y_\xi} \mathrm{B}^* f_1(\nu) \overline{\mathrm{B}^* f_2(\nu)}\, d\mathrm{Pl}_\xi(\nu) = \int_{\mathbb{R}^*} f_1(t) \overline{f_2(t)}\, \frac{dt}{|t|}.$$

This implies that B^* extends to a unitary injection

(5.19) $$L^2\left(\mathbb{R}^*, \tfrac{dt}{|t|}\right)_\xi \longrightarrow L^2\left(Y_\xi, d\mathrm{Pl}_\xi\right).$$

Let us substitute the expression for $\mathrm{B}^* f_1$ in Definition 5.2 into (5.18). There is no problem in changing the order of integration. Using the relation $k_\xi^{\varepsilon,\varepsilon'}(\bar{\nu}, t) = (\varepsilon\varepsilon')^\xi \overline{k_\xi^{\varepsilon,\varepsilon'}(\nu, t)}$ for ν in the support of $d\mathrm{Pl}_\xi$, we arrive at

(5.20) $$\int_{\mathbb{R}^*} f_1(t) \overline{\mathrm{BB}^* f_2(t)}\, \frac{dt}{|t|} = (\varepsilon\varepsilon')^\xi \int_{\mathbb{R}^*} f_1(t) \overline{f_2(t)}\, \frac{dt}{|t|}$$

for all $f_1, f_2 \in C_c^\infty (\mathbb{R}^*)_\xi$. This gives a right inverse of the transform $\mathrm{B} = \mathrm{B}_\xi^{\varepsilon,\varepsilon'}$:

PROPOSITION 5.3. *Let $\varepsilon, \varepsilon' \in \{1, -1\}$, $\xi \in \{0, 1\}$. For all $f \in C_c^\infty (\mathbb{R}^*)_\xi$:*

(5.21) $$\mathrm{B}_\xi^{\varepsilon,\varepsilon'} \left(\mathrm{B}_\xi^{\varepsilon,\varepsilon'}\right)^* f = (\varepsilon\varepsilon')^\xi f.$$

Let us write the right inverse $B^{-1} = (\varepsilon\varepsilon')^\xi B^*$ as

(5.22) $$\left(B_\xi^{\varepsilon,\varepsilon'}\right)^{-1} f(\nu) = 2(-\varepsilon\varepsilon')^\xi \int_0^\infty f(t) k_\xi^{\varepsilon,\varepsilon'}(\nu,t) \frac{dt}{t},$$

or more explicitly:

(5.23) $$\left(B_0^{\varepsilon,\varepsilon}\right)^{-1} f(\nu) = \int_0^\infty f(t) \frac{J_{-2\nu}(t) - J_{2\nu}(t)}{\sin \pi \nu} \frac{dt}{t},$$

$$\left(B_1^{\varepsilon,\varepsilon}\right)^{-1} f(\nu) = i\varepsilon \int_0^\infty f(t) \frac{J_{-2\nu}(t) + J_{2\nu}(t)}{\cos \pi \nu} \frac{dt}{t},$$

$$\left(B_0^{\varepsilon,-\varepsilon}\right)^{-1} f(\nu) = \frac{8}{\pi} \cos \pi\nu \int_0^\infty f(t) K_{2\nu}(t) \frac{dt}{t},$$

$$\left(B_1^{\varepsilon,-\varepsilon}\right)^{-1} f(\nu) = \frac{8i\varepsilon}{\pi} \sin \pi\nu \int_0^\infty f(t) K_{2\nu}(t) \frac{dt}{t},$$

and for $b \equiv \xi \mod 2$:

(5.24) $$\left(B_0^{\varepsilon,\varepsilon}\right)^{-1} f\left(\tfrac{b-1}{2}\right) = 2(-1)^{b/2} \int_0^\infty f(t) J_{b-1}(t) \frac{dt}{t},$$

$$\left(B_1^{\varepsilon,\varepsilon}\right)^{-1} f\left(\tfrac{b-1}{2}\right) = 2\varepsilon i^b \int_0^\infty f(t) J_{b-1}(t) \frac{dt}{t},$$

$$\left(B_\xi^{\varepsilon,-\varepsilon}\right)^{-1} f\left(\tfrac{b-1}{2}\right) = 0.$$

5.3. Vanishing delta term. Suppose that $\varepsilon = \varepsilon'$. If $f \in C_c^\infty(\mathbb{R}^*)_\xi$, the delta term in the sum formula leads to local integrals

(5.25) $$\int_{Y_\xi} B^* f(\nu) \, d\mathrm{Pl}_\xi(\nu).$$

With (3.49) and (5.4), we obtain:

(5.26) $$= 4\pi(-i\varepsilon)^\xi \int_{\mathrm{Re}\,\nu=0} \frac{\nu \, j^1 f(2\nu)}{\cos \pi\left(\nu + \tfrac{\xi}{2}\right)} \frac{d\nu}{2\pi i}$$

$$+ 2\varepsilon^\xi \sum_{b>1,\, b\equiv\xi \bmod 2} i^b (b-1) j^1 f(b-1).$$

Using estimate (5.6), we may shift the line of integration to the right to $\mathrm{Re}\,\nu = \sigma$, for arbitrarily large σ. This gives:

PROPOSITION 5.4. *Let $\varepsilon \in \{1,-1\}$, $\xi \in \{0,1\}$. For all $f \in C_c^\infty(\mathbb{R}^*)_\xi$:*

(5.27) $$\int_{Y_\xi} \left(B_\xi^{\varepsilon,\varepsilon}\right)^{-1} f \, d\mathrm{Pl}_\xi = 0.$$

Remark. There are elements $\varphi \in L^2\left(Y_\xi, d\mathrm{Pl}_\xi\right)$ for which $\int_{Y_\xi} \varphi \, d\mathrm{Pl}_\xi > 0$, see Lemma 3.18. This implies that the unitary map in (5.19) is not surjective.

6. Derivation of the Kloosterman sum formula

We return to the global context. We take Fourier term orders $r, r' \in \mathcal{O}' \setminus \{0\}$, $\varepsilon = \operatorname{sign} r$, $\varepsilon' = \operatorname{sign} r'$, and central character given by $\xi \in \{0, 1\}^d$.

Let us define the inverse Bessel transform

$$(6.1) \qquad \left(\mathrm{B}_\xi^{\varepsilon,\varepsilon'}\right)^{-1} f(v) := \prod_{j=1}^{d} (\varepsilon_j \varepsilon_j')^{\xi_j} \left(\mathrm{B}_{\xi_j}^{\varepsilon_j,\varepsilon_j'}\right)^* f_j(v_j)$$

for $v \in \mathbb{C}^d$, and f a function on $(\mathbb{R}^*)^d$ of product type $\bigotimes_{j=1}^d f_j$, satisfying $f_j(-t) = (\varepsilon_j \varepsilon_j')^{\xi_j} f_j(t)$. The integrals converge absolutely if $f_j \in C_c^\infty(\mathbb{R}^*)_{\xi_j}$. Thus, (6.1) defines a holomorphic function on $\mathbb{C}^d$. Lemma 5.1 shows that it is an element of $T_\xi^{\varepsilon\varepsilon'}(\tau, a)$ for all $\tau > 0$ and $a > 0$. Proposition 5.3 implies that

$$(6.2) \qquad \mathrm{B}_\xi^{\varepsilon,\varepsilon'} \left(\mathrm{B}_\xi^{\varepsilon,\varepsilon'}\right)^{-1} f = f$$

for such compactly supported smooth functions of product type. Although we denote it by $\left(\mathrm{B}_\xi^{\varepsilon,\varepsilon'}\right)^{-1}$, this operator is only a right inverse of $\mathrm{B}_\xi^{\varepsilon,\varepsilon'}$.

The parity condition on f associated to ξ implies that f is determined by its behavior on the connected component $(0, \infty)^d$ of $(\mathbb{R}^*)^d$. With the description (5.22), we are effectively working with functions on $(0, \infty)^d$.

Taking into account that the integrals in (5.25) vanish, we obtain from Theorem 3.21 the following formula valid for compactly supported test functions of product type:

PROPOSITION 6.1. *Let $r, r', \varepsilon, \varepsilon', \xi$ be as above, and let the central character determined by ξ be compatible with the character χ. Let f be a smooth compactly supported function of product type $f = \bigotimes_j f_j$, $f_j \in C_c^\infty(\mathbb{R}^*)$ satisfying the parity conditions $f_j(-t) = (\varepsilon_j \varepsilon_j')^{\xi_j} f_j(t)$. Then the sum of Kloosterman sums $\mathrm{K}_\chi^{r,r'}(f)$ given in (3.60) satisfies the following equality, with absolute convergence in both terms:*

$$(6.3) \qquad \mathrm{K}_\chi^{r,r'}(f) = \int_{Y_\xi} \left(\left(\mathrm{B}_\xi^{\varepsilon,\varepsilon'}\right)^{-1} f\right)(v) \, d\sigma_{\chi,\xi}^{r,r'}(v).$$

In this formula, we can prescribe the test function in the sum of Kloosterman sums $\mathrm{K}_\chi^{r,r'}(f)$. The conditions on f are rather strong: smooth, compactly supported, and of product type. We can use this formula to get estimates for sums of Kloosterman sums $S_\chi(r, r'; c)$, where $c = (c^{\sigma_1}, \ldots, c^{\sigma_d})$ runs through $I \setminus \{0\}$ intersected with rectangles in $\mathbb{R}^d$; see [7] for the case $\chi = 1$. However, it may be of interest to let c vary in regions of other type. In [8], regions of the form $|x_1 x_2| \leq X$, $A \leq |x_1/x_2| \leq B$ are used, in the real quadratic case $d = 2$. There, we employed general smooth compactly supported functions on $(0, \infty)^2$. Other applications also need more general test functions; e.g., the explicit formula for $|\zeta_F|^4$, see Motohashi, Theorem 4.2, [37], in the rational case.

6.1. Extension. We wish to have Proposition 6.1 for a less restricted class of test functions. The class we aim at will contain continuous functions that are not necessarily compactly supported, but have adequate decay as $t_j \downarrow 0, \to \infty$, for $1 \leq j \leq d$. For the extension, we follow the approach in §2 of [8].

Proposition 3.14 suggests that the following norms for functions on $(\mathbb{R}^*)^d$ are useful:

$$(6.4) \qquad N_{\alpha,\beta}(f) := \sup_{t \in (\mathbb{R}^*)^d} \frac{|f(t)|}{\prod_{j=1}^d \min\left(|t_j|^\alpha, |t_j|^{-\beta}\right)},$$

with $\alpha + \beta > 0$ and $\alpha > \frac{1}{2}$.

DEFINITION 6.2. For $\alpha, \beta \in \mathbb{R}$, let $\mathbf{F}_{\alpha,\beta}$ be the space of continuous functions f on $(\mathbb{R}^*)^d$ for which $N_{\alpha,\beta}(f) < \infty$. If $\xi \in \{0,1\}^d$, let $\mathbf{F}_{\alpha,\beta,\xi}$ be the subspace of $f \in \mathbf{F}_{\alpha,\beta}$ satisfying the parity conditions
$$f(t_1, \ldots, -t_j, \ldots, t_d) = (\varepsilon_j \varepsilon'_j)^{\xi_j} f(t_1, \ldots, t_j, \ldots, t_d)$$
for all $j = 1, \ldots, d$.

Proposition 3.14 shows that $f \mapsto K_\chi^{r,r'}(f)$ is a continuous linear form on $\mathbf{F}_{\alpha,\beta,\xi}$, for $\alpha > \frac{1}{2}, \alpha + \beta > 0$.

6.2. Estimates of inverse Bessel transforms. To study the continuity of the right hand side of (6.3) as a functional on $\mathbf{F}_{\alpha,\beta}$, we have to look carefully at estimates of $\varphi = \left(B_\xi^{\varepsilon,\varepsilon'}\right)^{-1} f$ on the support of $d\sigma_{\chi,\xi}^{r,r'}$.

As in the proof of Lemma 2 in [8], we have

$$(6.5) \qquad |\varphi(\nu)| \leq N_{\alpha,\beta}(f) \prod_{j=1}^{d} \left(I^0(\varepsilon_j, \varepsilon'_j, \xi_j; \nu) + I^\infty(\varepsilon_j, \varepsilon'_j, \xi_j; \nu) \right),$$

$$I^0(\varepsilon, \varepsilon', \delta; \nu) := \int_0^1 t^\alpha \left| k_\delta^{\varepsilon,\varepsilon'}(t, \nu) \right| \frac{dt}{t},$$

$$I^\infty(\varepsilon, \varepsilon', \delta; \nu) := \int_1^\infty t^{-\beta} \left| k_\delta^{\varepsilon,\varepsilon'}(t, \nu) \right| \frac{dt}{t},$$

with $\varepsilon, \varepsilon' \in \{1, -1\}$, $\delta \in \{0, 1\}$. The kernel functions $k_\delta^{\varepsilon,\varepsilon'}$ of the inverse Bessel transform are explicitly given in (5.23) and (5.24).

An advantage of (products of) $SL_2(\mathbb{R})$ over other Lie groups of real rank one is the wealth of results on Bessel functions that are available. Even for $SL_2(\mathbb{C})$ (where the kernel of the Bessel transform is a product of Bessel functions with complex argument) there is no ready substitute for Watson's book [50]. The results below will be needed later on, and, furthermore, they are useful in applications of the sum formula.

Power series expansion. The expansion (3.27) implies the estimate (3.63), which gives for $\operatorname{Re} \nu \geq -\frac{1}{4}$:

$$(6.6) \qquad \int_0^1 t^\alpha J_{2\nu}^{\pm 1}(t) \frac{dt}{t} \ll_\alpha \frac{1}{|\Gamma(2\nu + 1)|}.$$

This gives immediately

$$(6.7) \qquad I^0\left(\frac{b-1}{2}\right) \ll \frac{1}{(b-1)!} \qquad \text{for } \varepsilon\varepsilon' = 1, b \equiv \delta \bmod 2, b > 1,$$

$$(6.8) \qquad I^0(\nu) \ll |\nu|^{-1/2} \qquad \text{for } \operatorname{Re} \nu = 0, |\nu| \geq \frac{1}{4},$$

$$(6.9) \qquad I^0(\nu) \ll 1 \qquad \text{for } \operatorname{Re} \nu = 0, |\nu| \leq \frac{1}{4}, \delta = 1.$$

Here, and in the sequel, we omit $(\varepsilon, \varepsilon', \delta)$ from the formulas.

If $\delta = 0$, the zero of $\sin 2\pi\nu$ in the denominator in (3.35) is not canceled by the factor $\cos \pi\left(\nu + \frac{\delta}{2}\right)$. This case requires a bit more care. We have for $|\nu| = \frac{1}{4}, 0 < t \leq 1$:

$$(6.10) \qquad k_0^{\varepsilon,\varepsilon'}(\nu, t) = \frac{J_{-2\nu}^{\varepsilon\varepsilon'}(t) - J_{2\nu}^{\varepsilon\varepsilon'}(t)}{2 \sin \pi\nu} = O\left(t^{-1/4}\right).$$

	$I^0(\varepsilon,\varepsilon',\delta;\nu)$		$I^\infty(\varepsilon,\varepsilon',\delta;\nu)$	
	$\varepsilon\varepsilon'=1$	$\varepsilon\varepsilon'=-1$	$\varepsilon\varepsilon'=1$	$\varepsilon\varepsilon'=-1$
$\nu \in i\mathbb{R}$, $|\nu| \geq \frac{1}{4}$	$|\nu|^{-1/2}$ (6.8)		1 (6.18)	$|\nu|^{-\beta_0/2}$ (6.28)
$\nu \in i\mathbb{R}$, $|\nu| \leq \frac{1}{4}$, $\delta = 0$	1 (6.11)		1 (6.20)	1 (6.24)
$\delta = 1$	1 (6.9)		1 (6.19)	1 (6.24)
$0 < \nu \leq \frac{1}{4}$, $\delta = 0$	1 (6.11)		1 (6.21)	1 (6.25)
$\nu = \frac{b-1}{2}$, $b > 1$, $b \equiv \delta \bmod 2$	$\frac{1}{\Gamma(2\nu)}$ (6.7)	—	$\frac{1}{b}$ (6.13)	—

TABLE 2. Upper bounds for $I^0(\varepsilon,\varepsilon',\delta;\nu)$ and $I^\infty(\varepsilon,\varepsilon',\delta;\nu)$ of the type $O(A(\nu))$. In the entries in the table, we give $A(\nu)$ and a reference to the relevant formula. We assume $\alpha > \frac{1}{2}$ and $\beta > 0$. We take $\beta_0 = \beta$ if $\beta < 1$ and $\beta_0 = 1$ otherwise.

This estimate is uniform in ν and t; it extends to $|\nu| \leq \frac{1}{4}$, giving

$$(6.11) \qquad I^0(\nu) = O(1) \quad \text{for } |\nu| \leq \frac{1}{4},\ \delta = 0.$$

Bessel integral representation. The integral (3.66) gives for $\operatorname{Re}\nu > 0$:

$$(6.12) \qquad |J_\nu(t)| \ll \left(1 + \frac{1}{\operatorname{Re}\nu}\right) e^{\frac{\pi}{2}|\operatorname{Im}\nu|}.$$

Hence, under the assumption $\beta > 0$:

$$(6.13) \qquad I^\infty\left(\tfrac{b-1}{2}\right) \ll \frac{1}{b} \quad \text{for } \varepsilon\varepsilon' = 1,\ b \equiv \delta \bmod 2,\ b > 1.$$

Mellin integral for $J_{2\nu}$. Deshouillers and Iwaniec, have shown several other ways to estimate Bessel functions; see §7.2 of [**12**]. Here we use

$$(6.14) \qquad J_{2\nu}(t) = \frac{1}{4\pi i}\int_{\operatorname{Re} s = \sigma} \left(\frac{t}{2}\right)^{-s} \frac{\Gamma\left(\nu + \frac{s}{2}\right)}{\Gamma\left(1 + \nu - \frac{s}{2}\right)} ds + \frac{(t/2)^{2\nu}}{\Gamma(2\nu+1)}$$

with $-1 < \operatorname{Re}\nu + \frac{\sigma}{2} < 0$ and $\sigma < 0$. (One can relate this integral to the power series expansion by moving the line of integration to the left.)

The integral will give a factor $t^{-\sigma}$ in the estimate. Let us take σ slightly smaller than 0, and $\sigma > -\beta$, to avoid convergence problems in the integrals I^∞ in (6.5). Now we can apply (6.14) for $-\frac{1}{4} \leq \operatorname{Re}\nu \leq \frac{-\sigma}{4}$.

6. DERIVATION OF THE KLOOSTERMAN SUM FORMULA

First we estimate the integral. In the resulting estimate there is a factor $t^{-\sigma}$ and an integral

$$\int_{-\infty}^{\infty} \frac{e^{-\frac{\pi}{2}|\operatorname{Im} \nu + \frac{y}{2}| + \frac{\pi}{2}|\operatorname{Im} \nu - \frac{y}{2}|}}{\left(1 + \left|\operatorname{Im} \nu + \frac{y}{2}\right|\right)^{\frac{1}{2} - \operatorname{Re} \nu - \frac{\sigma}{2}} \left(1 + \left|\operatorname{Im} \nu - \frac{y}{2}\right|\right)^{\frac{1}{2} + \operatorname{Re} \nu - \frac{\sigma}{2}}} \, dy.$$

We write $a = |\operatorname{Im} \nu|$, $\pm 1 = \operatorname{sign} \operatorname{Im} \nu$, and $b = \operatorname{Re} \nu$. The integral over $y \geq 2a$ is estimated by

(6.15)
$$\int_{2a}^{\infty} \frac{e^{-\pi \operatorname{Im} \nu}}{(1 \pm a + y/2)^{\frac{1}{2} - b - \frac{\sigma}{2}} (1 \mp a + y/2)^{\frac{1}{2} + b - \frac{\sigma}{2}}} \, dy$$

$$\ll e^{-\pi \operatorname{Im} \nu} \int_{2a}^{3a} \frac{dy}{(1+a)^{\frac{1}{2} \mp b - \frac{\sigma}{2}} (1 - a + y/2)^{\frac{1}{2} \pm b - \frac{\sigma}{2}}}$$

$$+ e^{-\pi \operatorname{Im} \nu} \int_{3a}^{\infty} \frac{dy}{(1 + y/2)^{1-\sigma}}$$

$$\ll (1+a)^{\sigma} e^{-\pi \operatorname{Im} \nu}.$$

In a similar way, the integral over $y \leq -2a$ is $O\left((1+a)^{\sigma} e^{\pi \operatorname{Im} \nu}\right)$.

In the range $-2a \leq y \leq 2a$, the numerator of the integrand is $e^{\mp \frac{\pi}{2} y}$. The denominator is larger than $(1+a)^{\frac{1}{2} - |b| - \frac{\sigma}{2}}$. This leads to an estimate by $e^{\pi |\operatorname{Im} \nu|} (1+a)^{-\frac{1}{2} + |b| + \frac{\sigma}{2}}$. We have taken σ near 0. So we have obtained for ν in the region $-\frac{1}{4} \leq \operatorname{Re} \nu \leq \frac{-\sigma}{2}$:

(6.16) $$J_{2\nu}(t) \ll t^{-\sigma} e^{\pi |\operatorname{Im} \nu|} (1 + |\operatorname{Im} \nu|)^{\sigma}.$$

Combining this with (6.12) for $\operatorname{Re} \nu \geq \frac{-\sigma}{2}$, we obtain, uniformly for $t > 0$ and $|\operatorname{Re} \nu| \leq \frac{1}{4}$:

(6.17) $$J_{2\nu}(t) \ll t^{-\sigma} e^{\pi |\operatorname{Im} \nu|}.$$

This implies, for $\beta > 0$:

(6.18) $$I^{\infty}(\nu) \ll_{\beta} 1 \quad \text{for } \operatorname{Re} \nu = 0, \; \varepsilon \varepsilon' = 1, \; |\nu| \geq \frac{1}{4},$$

(6.19) $$\quad \text{for } \operatorname{Re} \nu = 0, \; \varepsilon \varepsilon' = 1, \; \delta = 1,$$

Consider now $\delta = 0$. On $|\nu| = \frac{1}{4}$, we have $k_0^{\varepsilon,\varepsilon}(\nu, t) = O(t^{-\sigma})$ for $t \geq 1$. This extends to $|\nu| \leq \frac{1}{4}$, and gives, for $\beta > 0$:

(6.20) $$I^{\infty}(\nu) \ll_{\beta} 1 \quad \text{for } \operatorname{Re} \nu = 0, \; \varepsilon \varepsilon' = 1, \; \delta = 0, \; |\nu| \leq \frac{1}{4},$$

(6.21) $$\quad \text{for } 0 < \nu \leq \frac{1}{4}, \; \varepsilon \varepsilon' = 1, \; \delta = 0.$$

Basset's integral. Partial integration applied to (3.67) yields

(6.22) $$K_{2\nu}(t)$$
$$= \frac{i}{\sqrt{\pi}} \Gamma\left(2\nu + \tfrac{3}{2}\right) 2^{2\nu} t^{-2\nu - 1} \int_{-\infty}^{\infty} e^{-itu} \left(u^2 + 1\right)^{-2\nu - \frac{3}{2}} u \, du$$

for $\operatorname{Re} \nu > -\frac{1}{4}$. Hence, for $\beta > -1 - 2\operatorname{Re}\nu$:

(6.23) $$K_{2\nu}(t) \ll \frac{(1 + |\operatorname{Im}\nu|)^{1+2\operatorname{Re}\nu}}{e^{\pi|\operatorname{Im}\nu|}t^{1+2\operatorname{Re}\nu}} \quad \text{for } t > 0, \operatorname{Re}\nu \geq 0,$$

(6.24) $$I^\infty(\nu) \ll 1 + |\nu| \quad \text{if } \varepsilon\varepsilon' = -1, \operatorname{Re}\nu = 0,$$

(6.25) $$\ll 1 \quad \text{if } \varepsilon\varepsilon' = -1, 0 < \nu < \frac{1}{2}.$$

If $|\nu| \geq \frac{1}{4}$, the estimate (6.24) is inferior in the ν-aspect to the other results up till now. One can do better:

Mellin integral for $K_{2\nu}$. For $\operatorname{Re}\nu = 0$, $|\nu| \geq \frac{1}{4}$, we apply the representation

(6.26) $$K_{2\nu}(t) = \frac{1}{2\pi i} \int_{\operatorname{Re}\nu = \sigma} 2^{s-2} \Gamma\left(\frac{s}{2} + \nu\right) \Gamma\left(\frac{s}{2} - \nu\right) t^{-s} ds$$
$$+ 4^{-1+\nu} \Gamma(2\nu) t^{-2\nu} + 4^{-1-\nu} \Gamma(-2\nu) t^{2\nu},$$

with $-1 < \sigma < 0$. Estimating this integral is easier than that in (6.14). It leads to

(6.27) $$K_{2\nu}(t) \ll e^{-\pi|\operatorname{Im}\nu|} \left(t^{-\sigma}(1 + |\nu|)^\sigma + (1 + |\nu|)^{-1/2}\right).$$

We assume $\beta > 0$, and take $\sigma = \max\left(-\frac{\beta}{2}, -\frac{1}{2}\right)$, to obtain

(6.28) $$I^\infty(\nu) \ll (1 + |\nu|)^\sigma.$$

Recapitulation. Table 2 gives an overview of the estimates for $I^0(\nu)$ and $I^\infty(\nu)$. In the light of these estimates, we can now replace (6.5) by

(6.29) $$\left(\left(B_\xi^{\varepsilon,\varepsilon'}\right)^{-1} f\right)(\nu) \ll_{\alpha,\beta} N_{\alpha,\beta}(f) \prod_{j=1}^d (1 + |\nu_j|)^{-\min(1,\beta)/2}$$

for $\nu \in \operatorname{Supp} d\sigma_{\chi,\xi}^{r,r'}$, $f \in \mathbf{F}_{\alpha,\beta,\xi}$, $\alpha > \frac{1}{2}, \beta > 0$.

However, this result is not good enough for integrability with respect to $d\sigma_{\chi,\xi}^{r,r'}$.

6.3. Differentiation. The local kernels $k_{\xi_j}^{\varepsilon_j,\varepsilon_j'}(\nu, t)$, see (5.23), are eigenfunctions of the differential operator $\mathcal{B}^{\varepsilon_j \varepsilon_j'}$ given by

(6.30) $$\mathcal{B}^{\pm 1} := t^2 \partial_t^2 + t \partial_t \pm t^2,$$

with eigenvalue $4\nu^2$. This opens the possibility to improve (6.29) by partial integration.

The operator $\mathcal{B}^{\pm 1}$ is its own transpose:

(6.31) $$\int_0^\infty \mathcal{B}^{\pm 1} f(t) \, g(t) \frac{dt}{t} = \int_0^\infty f(t) \, \mathcal{B}^{\pm 1} g(t) \frac{dt}{t}$$

for C^2-functions f and g with sufficient decay near 0 and ∞.

We take $g(t) = k_{\xi_j}^{\varepsilon_j,\varepsilon_j'}(t, \nu)$. We work with a fixed $\nu \in i\mathbb{R} \cup \left(0, \frac{1}{4}\right] \cup \left(\frac{1}{2} + \mathbb{N}_0\right)$ if $\xi_j = 0$, $\nu \in i\mathbb{R} \cup \mathbb{N}$ if $\xi_j = 1$. The behavior near 0 of $g^{(l)}(t)$, $0 \leq l \leq 2$, can be derived from the power series expansion (3.27).

Near ∞, we use the asymptotic expansions of Bessel functions. In the case $\varepsilon\varepsilon' = -1$, we find exponential decay of $K_{2\nu}(t)$ in 7.23 (1) in [50]. The relation $K'_{2\nu}(t) = -\frac{1}{2}K_{2\nu+1}(t) -$

$\frac{1}{2}K_{2\nu-1}(t)$ shows that the derivative decays exponentially too; see 3.71 (2) in [50]. For $\varepsilon_j \varepsilon'_j = 1$, we use

(6.32) $$J_{2\nu}(t) = \sqrt{\frac{2}{\pi t}} \left(\cos\left(t - \pi\nu - \tfrac{\pi}{4}\right) + O\left(t^{-1}\right) \right) \qquad (t \to \infty);$$

see [50], 7.21 (1). With $J'_{2\nu}(t) = \frac{1}{2}J_{2\nu-1}(t) - \frac{1}{2}J_{2\nu+1}(t)$ (see 3.2 (2), [50]), this gives $O(t^{-1/2})$ for the derivative. The behavior of the second derivative can be derived from the Bessel differential equation. Taking the cases $\varepsilon_j \varepsilon'_j = 1$ and $\varepsilon_j \varepsilon'_j = -1$ together, we obtain the following:

(6.33)

	near 0			near ∞
	$\nu \in Y_\xi$	$\nu = \frac{b-1}{2}$		
$g(t)$	$O\left(\lvert t\rvert^{-2\lvert\operatorname{Re}\nu\rvert}\right)$	$O\left(t^{b-1}\right)$		$O\left(t^{-1/2}\right)$
$tg'(t)$	$O\left(\lvert t\rvert^{-2\lvert\operatorname{Re}\nu\rvert}\right)$	$O\left(t^{b-1}\right)$		$O\left(t^{1/2}\right)$
$t^2 g''(t)$	$O\left(\lvert t\rvert^{-2\lvert\operatorname{Re}\nu\rvert}\right)$	$O\left(t^{b-1}\right)$		$O\left(t^{3/2}\right)$

For $\nu = \frac{b-1}{2}$, $b \geq 1$, $b \equiv \xi \bmod 2$, the estimate $O\left(\lvert t\rvert^{b-1}\right)$ as $\lvert t\rvert \downarrow 0$ is better than $O\left(\lvert t\rvert^{-2\lvert\operatorname{Re}\nu\rvert}\right) = O\left(\lvert t\rvert^{1-b}\right)$.

Let us write $D = t\partial_t$. We look for conditions

$$D^l f(t) \ll \begin{cases} t^{\alpha_l} & \text{as } t \downarrow 0, \\ t^{-\beta_l} & \text{as } t \to \infty, \end{cases} \qquad (0 \leq l \leq 2),$$

that ensure that (6.31) can be applied. We need integrability against $\frac{dt}{t}$ of $t^2 fg$, $(D^2 f)g$, $(Df)(Dg)$, and $f(D^2 g)$, and the vanishing of $(Df)g$ and $f(Dg)$ at the boundary. Near 0, the conditions are $\alpha_l > 2\lvert\operatorname{Re}\nu\rvert$ if $\nu \in i\mathbb{R} \cup \left(0, \tfrac{1}{4}\right]$, and $\alpha_l > 1 - b$ if $\nu = \frac{b-1}{2} > \frac{1}{4}$. We can take any $\alpha_l > \frac{1}{2}$. Near ∞, it suffices to take $\beta_0 > \frac{3}{2}, \beta_1 > \frac{1}{2}, \beta_2 > -\frac{1}{2}$. Under these conditions, a twice continuously differentiable f on $(0, \infty)$ satisfies

(6.34) $$\int_0^\infty \left(\mathcal{B}^{\pm 1} f\right)(t) k_\xi^{\varepsilon_j, \varepsilon'_j}(t, \nu) \, \frac{dt}{t} = 4\nu^2 \int_0^\infty f(t) k_\xi^{\varepsilon_j, \varepsilon'_j}(t, \nu) \, \frac{dt}{t}.$$

This can be applied to functions on $(0, \infty)^d$, with respect to each of the variables. Let $D_j := t_j \partial_{t_j}$ and $\mathcal{B}_j^{\pm 1} := D_j^2 \pm t_j^2$.

DEFINITION 6.3. For $\alpha > \frac{1}{2}$ and $\beta > \frac{3}{2}$, we define $\mathbf{F}_{\alpha,\beta}^{(1)}$ as the subspace of sufficiently differentiable $f \in \mathbf{F}_{\alpha,\beta}$ (see Definition 6.2) such that the derivatives $D_1^{n_1} \cdots D_d^{n_d} f$ are continuous for all $(n_1, \ldots, n_d) \in \{0, 1, 2\}^d$ and satisfy

(6.35) $$D_1^{n_1} \cdots D_d^{n_d} f(t) \ll_f \prod_{j=1}^d \min\left(t_j^\alpha, t_j^{-(\beta - n_j)}\right).$$

Let $\mathbf{F}_{\alpha,\beta,\xi}^{(1)} := \mathbf{F}_{\alpha,\beta}^{(1)} \cap \mathbf{F}_{\alpha,\beta,\xi}$.

For $f \in \mathbf{F}_{\alpha,\beta}^{(1)}$, partial integration is possible in all coordinates. So we have, for $(m_1, \ldots, m_d) \in \{0,1\}^d$, $v \in Y_\xi$:

(6.36)
$$\left(\mathcal{B}_\xi^{\varepsilon,\varepsilon'}\right)^* \left(\left(\mathcal{B}_1^{\varepsilon_1 \varepsilon'_1}\right)^{m_1} \cdots \left(\mathcal{B}_d^{\varepsilon_d \varepsilon'_d}\right)^{m_d} f\right)(v)$$
$$= \int_{(0,\infty)^d} \left(\left(\mathcal{B}_1^{\varepsilon_1 \varepsilon'_1}\right)^{m_1} \cdots \left(\mathcal{B}_d^{\varepsilon_d \varepsilon'_d}\right)^{m_d} f\right)(t) \prod_{j=1}^{d} k_{\xi_j}^{\varepsilon_j,\varepsilon'_j}(t_j, v_j) \prod_{j=1}^{d} \frac{dt_j}{t_j}$$
$$= \int_{(0,\infty)^d} f(t) \prod_{j=1}^{d} \left(4v_j^2 k_{\xi_j}^{\varepsilon_j,\varepsilon'_j}(t_j, v_j)\right) \prod_{j=1}^{d} \frac{dt_j}{t_j}$$
$$= \prod_{j=1}^{d}(4v_j^2) \left(\mathcal{B}_\xi^{\varepsilon,\varepsilon'}\right)^* f(v).$$

Let us put, for a given $v \in Y_\xi \setminus \left(\frac{1}{4}, \frac{1}{2}\right)$:

$$\gamma_j := \begin{cases} 1 & \text{if } v_j^2 > 0, \\ -1 & \text{if } v_j^2 < 0, \\ 0 & \text{if } v_j = 0. \end{cases}$$

Then
$$\tilde{f} = \left(1 + \gamma_1 \mathcal{B}_1^{\varepsilon_1 \varepsilon'_1}\right)\left(1 + \gamma_2 \mathcal{B}_2^{\varepsilon_2 \varepsilon'_2}\right) \cdots \left(1 + \gamma_d \mathcal{B}_d^{\varepsilon_d \varepsilon'_d}\right) f$$

is an element of $\mathbf{F}_{\alpha,\beta-2}$ if $f \in \mathbf{F}_{\alpha,\beta}^{(1)}$, and

(6.37)
$$\left(\left(\mathbf{B}_\xi^{\varepsilon,\varepsilon'}\right)^{-1} \tilde{f}\right)(v) = \prod_{j=1}^{d}\left(1 + 4|v_j|^2\right) \left(\mathbf{B}_\xi^{\varepsilon,\varepsilon'}\right)^{-1} f(v).$$

To obtain an improvement of (6.29), we assume $\alpha > \frac{1}{2}, \beta > 2$. We apply (6.29) to $\tilde{f}$:

(6.38)
$$\prod_{j=1}^{d}\left(1 + 4|v_j|^2\right) \left(\left(\mathbf{B}_\xi^{\varepsilon,\varepsilon'}\right)^{-1} f\right)(v) = \left(\mathbf{B}_\xi^{\varepsilon,\varepsilon'}\right)^{-1} \tilde{f}(v)$$
$$\ll N_{\alpha,\beta-2}(\tilde{f}) \prod_{j=1}^{d}\left(1 + |v_j|\right)^{-\delta}$$

with $\delta > 0$. Hence

(6.39)
$$\left(\mathbf{B}_\xi^{\varepsilon,\varepsilon'}\right)^{-1} f(v) \ll N_{\alpha,\beta-2}(\tilde{f}) \prod_{j=1}^{d}\left(1 + |v_j|\right)^{-b-2}.$$

The norm $N_{\alpha,\beta-2}(\tilde{f})$ is estimated by the following norm $N_{\alpha,\beta}^{(1)}(f)$ for $f \in \mathbf{F}_{\alpha,\beta}^{(1)}$:

(6.40)
$$N_{\alpha,\beta}^{(1)}(f) := \max_{(n_1,\ldots,n_d)} \sup_t \frac{|D_1^{n_1} \cdots D_d^{n_d} f(t)|}{\prod_{j=1}^{d} \min\left(t_j^\alpha, t_j^{-(\beta-n_j)}\right)},$$

with $(n_1, \ldots, n_d)$ running through $\{0,1,2\}^d$, and t running through $(0,\infty)^d$.

PROPOSITION 6.4. *Let* $\alpha > \frac{1}{2}, \beta > 2$. *There exists* $a > 2$ *such that for all* $f \in \mathbf{F}_{\alpha,\beta,\xi}^{(1)}$:

(6.41)
$$\left(\left(\mathbf{B}_\xi^{\varepsilon,\varepsilon'}\right)^{-1} f\right)(v) \ll_{\alpha,\beta} N_{\alpha,\beta}^{(1)}(f) \prod_{j=1}^{d}\left(1 + |v_j|\right)^{-a} \quad (v \in Y_\xi),$$

uniformly for $v \in \operatorname{Supp} d\sigma_{\chi,\xi}^{r,r'}$.

Thus we have $\left(\mathbf{B}_{\xi}^{\varepsilon,\varepsilon'}\right)^{-1} \varphi(v) \ll \varphi_{a,p}(v)$ for $v \in Y_{\xi}$, with $\varphi_{a,p}$ as defined in Lemma 3.18, $\frac{1}{2} < p < 1$. This shows that $\left(\mathbf{B}_{\xi}^{\varepsilon,\varepsilon'}\right)^{-1} f$ is integrable for $d\sigma_{\chi,\xi}^{r,r'}$.

6.4. Extension to compactly supported functions. Let us consider two continuous linear forms on $F_{\alpha,\beta,\xi}^{(1)}$, with $\alpha > \frac{1}{2}, \beta > 2$:

(6.42)
$$k : f \mapsto \mathrm{K}_{\chi}^{r,r'}(f),$$

$$s : f \mapsto \int_{Y_{\xi}} \left(\left(\mathbf{B}_{\xi}^{\varepsilon,\varepsilon'}\right)^{-1} f\right)(v) \, d\sigma_{\chi,\xi}^{r,r'}(v).$$

The functions in $\mathbf{F}_{\alpha,\beta,\xi}^{(1)}$ satisfy

(6.43)
$$f(t_1, \ldots, -t_j, \ldots, t_d) = (\varepsilon_j \varepsilon_j')^{\xi_j} f(t_1, \ldots, t_d) \qquad (1 \leq j \leq d).$$

This symmetry condition allows us to consider k and s as distributions on $(0, \infty)^d$. The order of k is zero, the order of s is at most $2d$, see Proposition 6.4. Proposition 6.1 gives equality of k and s on $\bigotimes_{j=1}^{d} C_c^{\infty}(0, \infty)$. This implies that k and s coincide on $C_c^{\infty}\left((0, \infty)^d\right)$; see the proof of Theorem 5.1.1, Chap. V, [19]. This shows that the sum formula (6.3) holds for all $f \in C_c^{\infty}\left((\mathbb{R}^*)^d\right)$ satisfying (6.43).

6.5. Further extension. Now we are in a position to prove the second main result in this paper, the *Kloosterman sum formula*:

THEOREM 6.5. *Let* $r, r' \in \mathcal{O}' \setminus \{0\}$, *and let* $\xi \in \{0, 1\}^d$ *determine a central character compatible with* χ. *Put* $\varepsilon = \operatorname{sign} r$, $\varepsilon' = \operatorname{sign}(r')$.

Let $\alpha > \frac{1}{2}, \beta > 2$. *For all* $f \in \mathbf{F}_{\alpha,\beta,\xi}^{(1)}$, *see Definition 6.3, the following equality holds, with absolute convergence of the integral and the sum of Kloosterman sums:*

(6.44)
$$\sum_{c \in I \setminus \{0\}} \frac{S_{\chi}(r, r'; c)}{|N(c)|} f\left(\tfrac{4\pi |rr'|^{1/2}}{c}\right) = \int_{Y_{\xi}} \left(\left(\mathbf{B}_{\xi}^{\varepsilon,\varepsilon'}\right)^{-1} f\right)(v) \, d\sigma_{\chi,\xi}^{r,r'}(v).$$

The space $\mathbf{F}_{\alpha,\beta}^{(1)}$ contains all compactly supported functions that satisfy the parity condition (6.43). It is the space of all functions on $(\mathbb{R}^*)^d$ satisfying (6.43) and the conditions in Definitions 6.2 and 6.3.

By $\frac{4\pi |rr'|^{1/2}}{c}$ we mean $\left(\frac{4\pi\sqrt{|r_1 r_1'|}}{c_1}, \ldots, \frac{4\pi\sqrt{|r_d r_d'|}}{c_d}\right) \in (\mathbb{R}^*)^d$. The measure $d\sigma_{\chi,\xi}^{r,r'}$ has been introduced in (3.43); it contains products of Fourier coefficients of automorphic forms.

PROOF. Take $\alpha_1 \in \left(\frac{1}{2}, \alpha\right), \beta_1 \in (2, \beta)$. The continuous linear forms k and s on $\mathbf{F}_{\alpha_1,\beta_1,\xi}^{(1)}$, see (6.42), are equal on the subspace of compactly supported functions.

Let χ be a cut-off function of the form

$$\chi(t_1, \ldots, t_d) = \prod_{j=1}^{d} \chi_X\left(\log |t_j|\right),$$

with $\chi_X \in C_c^\infty(\mathbb{R})$, $\chi_X = 1$ on $[-X, X]$ and $\chi = 0$ outside $[-X-1, X+1]$, such that χ_X' and χ_X'' are bounded uniformly in $X \geq 1$. We find for $f \in \mathbf{F}_{\alpha,\beta,\xi}^{(1)} \subset \mathbf{F}_{\alpha_1,\beta_1,\xi}^{(1)}$ uniformly in $X \geq 1$:

$$N_{\alpha_1,\beta_1}^{(1)}(f - \chi f) \ll \max_{n_1,\ldots,n_d} \sup_{t, |\log|t_j|| \geq X} \frac{|D^{n_1} \cdots D^{n_d} f(t)|}{\prod_{j=1}^d \min\left(|t_j|^{\alpha_1}, |t_j|^{-\beta_1+\eta_j}\right)}$$

$$\ll N_{\alpha,\beta}^{(1)}(f) \prod_{j=1}^d \min\left(e^{-X(\alpha-\alpha_1)}, e^{X(\beta_1-\beta)}\right)$$

(6.45) $$= o\left(N_{\alpha,\beta}^{(1)}(f)\right) \quad \text{as } X \to \infty.$$

So $\mathbf{F}_{\alpha,\beta,\xi}^{(1)}$ is contained in the closure of $C_c^\infty\left((\mathbb{R}^*)^d\right) \cap F_{\alpha_1,\beta_1,\xi}^{(1)}$ in $\mathbf{F}_{\alpha,\beta,\xi}^{(1)}$. Hence the equality of k and s in (6.42) extends to the space $\mathbf{F}_{\alpha,\beta,\xi}^{(1)}$. □

Comparison. We compare the class of test functions in the Kloosterman sum formula, for the case $d = 1$, with those given by Kuznetsov and Proskurin. For the case $\xi = 0$, the test function is even, and determined by its values on $(0, \infty)$.

Teorema 2 in [29]	*Teorema* on p. 33 of [41]	Theorem 6.44 here
$f \in C^3([0, \infty))$		$f \in C^2(0, \infty)$
Behavior near 0		
$f(0) = f'(0) = 0$		$f^{(k)}(t) = O(t^{\alpha-k})$
		for $0 \leq k \leq 2$, $\alpha > \frac{1}{2}$
Behavior near ∞		
$f(x) \ll x^{-B}$	$f(x) \ll x^{-1-\varepsilon}$	$f(x) \ll x^{-\beta}$
$f'(x) \ll x^{-B}$	$f'(x) \ll x^{-2-\varepsilon}$	$f'(x) \ll x^{-\beta}$
$f''(x) \ll x^{-B}$	$f''(x) \ll x^{-2-\varepsilon}$	$f''(x) \ll x^{-\beta}$
$f'''(x) \ll x^{-B}$	$f'''(x) \ll x^{-2-\varepsilon}$	with $\beta > 2$
with $B > 2$	with $\varepsilon > 0$	

In most applications, the difference between these classes of test functions is not likely to be important.

7. Application

Theorem 6.5 is suitable to estimate sums of Kloosterman sums. In the case $F = \mathbb{Q}$, $I = \mathbb{Z}$, Kuznetsov, [29], has shown that for $n, m \geq 1$

(7.1) $$\sum_{c=1}^X \frac{S(n, m; c)}{c} = O\left(X^{1/6} (\log X)^{1/3}\right) \quad (X \to \infty).$$

Using solely the Weil estimate, (2.37), one cannot do better than $O\left(X^{1/2+\varepsilon}\right)$ for this sum. The sum formula shows that the cancellation is considerable.

In the Hilbert modular context, sums of Kloosterman sums have been estimated in [7], for the case $\xi = 0$, $\chi = 1$. To extend these results to the present more general situation requires work that we have not carried out completely. We do believe that by the methods

7. APPLICATION

in [7] and using the sum formula in Theorem 6.5, one should obtain (similar) non trivial estimates of sums of Kloosterman sums in the present context.

We mention some steps that are important in [7]:

- We estimate the sum

(7.2) $$\Lambda(\mathbf{X}) = \sum_{c \in I \setminus \{0\},\, \frac{1}{2} X_j < |c_j| \leq X_j} \frac{S(r, r'; c)}{|N(c)|}$$

for various combinations of the $X_j > 0$. Note that although $|N(c)| \geq 1$, some of the coordinates of c may get close to zero. So small and large X_j have to be considered.

Without the sum formula, only based on the Weil type estimate (2.47) of Kloosterman sums, Lemma 4.1.2 in [7] implies for each $\varepsilon > 0$:

(7.3) $$\Lambda(\mathbf{X}) \ll_{F,r,r',\varepsilon} \prod_{j=1}^{d} \left(X_j^{\varepsilon} \max\left(X_j^{1/2}, X_j^{-1/2} \right) \right).$$

- The sum in (7.2) is compared with the sum with smooth bounds

(7.4) $$\sum_{c \in I \setminus \{0\}} \psi(c) \frac{S(r, r'; c)}{|N(c)|},$$

where $\psi \in C_c^{\infty}(\mathbb{R}^d)$ is a smooth approximation of the box $\frac{1}{2} \leq \frac{1}{2} X_j \leq |c_j| \leq X_j$. The difference between this smoothly bounded sum and the sharply bounded sum in (7.2) is estimated with the Weil type estimate (2.47); see Lemma 4.1.2 in [7].

- The Kloosterman sum formula transforms the sum (7.4) into

(7.5) $$\int_{Y_\xi} b(v)\, d\sigma_{\chi,\xi}^{r,r'}(v),$$

where b is obtained from ψ by Bessel transformation.

- Use results on Bessel functions to get good estimates of $b(v)$ on various regions in Y_ξ.
- Use the spectral sum formula to get an estimate of

$$\int_{v \in Y_\xi,\, |v_j| \leq N_j} \left| d\sigma_{\chi,\xi}^{r,r'} \right|$$

for large N_j. See Corollary 3.3.2 in [7]. This result may be viewed as a precursor of the density results in §4.

- Combine this to obtain an estimate of the integral in (7.5), and then of the sum of Kloosterman sums (7.2). Proposition 4.6.1 in [7] implies, under the assumption that there are no exceptional eigenvalues at all real places, for each $\varepsilon > 0$, with $X_j \geq 1$ for all j:

(7.6) $$\Lambda(\mathbf{X}) = O_{F,I,r,r',\varepsilon}\left(\frac{\prod_{j=1}^{d} X_j^{1/2+\varepsilon}}{\min_{1 \leq j \leq d} X_j} \right) + O_{F,I,r,r',\varepsilon}\left(\prod_{j=1}^{d} X_j^{\frac{d}{2(d+2)}+\varepsilon} \right).$$

- Finally, work has to be done to arrive from (7.3) and (7.6) at an estimate of sums of Kloosterman sums in which c runs over a large cube in $\mathbb{R}^d$ centered at the origin. For $d \geq 2$, in the absence of exceptional eigenvalues, the result in

Corollary 4.7.2 in [**7**] is

(7.7) $$\sum_{c\in I\setminus\{0\},\, \forall_j |c_j|\leq X} \frac{S(r,r';c)}{|N(c)|} \ll_{F,I,t,t',\varepsilon} X^{\frac{d-1}{2}+\varepsilon} \quad (X\to\infty).$$

For $F = \mathbb{Q}$, Kuznetsov's result (7.1) can be recovered, and, in the absence of exceptional eigenvalues, similar estimates for non-zero ideals in $\mathbb{Z}$, see (79) on p. 148 of [**7**], and the discussion following it.

8. Final comments

In the approach to the sum formulas in this paper, the principal objects on the spectral side are automorphic representations, not individual automorphic forms. The approach is via a scalar product of Poincaré series, or the closely related approach via a two-sided Fourier coefficient of a kernel function. First the spectral sum formula is proved. If a (right) inverse of the "Bessel transform" is available, the Kloosterman sum formula follows.

We mention a selection of papers that follow this approach: Bruggeman, [**4**], considers automorphic forms on the universal covering group of $SL_2(\mathbb{R})$; this leads to a sum formula for all real weights. Bruggeman, Miatello and Pacharoni treat, in [**7**], automorphic forms on SL_2 over a totally real number field. Imaginary quadratic number fields are considered by Bruggeman and Motohashi, [**10**] (PSL_2 over $\mathbb{Q}(i)$), and by Lokvenec-Guleska, [**33**] (SL_2 over any quadratic number field). Venkatesh, [**47**], gives an adelic sum formula (GL_2 over any number field, K-trivial case).

The ingredients to obtain both versions of the sum formula over SL_2 for arbitrary number fields are available, see [**10**], [**33**], [**47**]. However, carrying out the proof in detail involves many technicalities, and seems quite a time consuming job.

Cogdell and Piatetski-Shapiro, [**11**], show that one may start directly with the decomposition of $L^2(SL_2(\mathbb{Z})\backslash SL_2(\mathbb{R}))$ into irreducible subspaces. They arrive directly at the Kloosterman sum formula.

In general, one can envisage a sum formula for $\Gamma\backslash G$ where Γ is a cofinite discrete subgroup of a Lie group G. Miatello and Wallach, [**35**], prove a sum formula where G is any Lie group of real rank one in the so called spherical case; that is, only automorphic representations that have non-zero K-fixed vectors occur. In [**51**] the same authors consider products of rank one groups, for a special class of test functions. For a general Lie group G of real rank one, it is not known how to invert the "Bessel transform" generalizing (1.8).

The main applications of the sum formulas discussed above give results concerning weighted densities of spectral data and sums of Kloosterman sums. For this purpose a precise understanding of the Bessel transform is desirable. There is an important related direction of work, in which automorphic data for different groups are related, for instance GL_2 over a number field and over a quadratic extension. Using a relative "Kloosterman integral" over a quadratic extension leads to relations between automorphic representations for both groups. For the relative trace formula and the related Kloosterman integrals, we point the reader to [**53**], [**23**], [**54**], [**22**], and the references therein.

APPENDIX A

Sum formula for the congruence subgroup $\Gamma_1(I)$

We have derived Theorems 3.21 and 6.5 for $\Gamma = \Gamma_0(I)$ with a character $\chi : \begin{pmatrix} a & b \\ c & d \end{pmatrix} \mapsto \chi(d)$. The group $\Gamma_1(I)$ is a normal subgroup of $\Gamma = \Gamma_0(I)$, with quotient isomorphic to $(O/I)^*$; see §2.1.1. So the space $L^2(\Gamma_1(I)\backslash G)$ is the direct sum of spaces $L^2(\Gamma_0(I)\backslash G, \chi)$ where χ runs over the characters χ that are trivial on $\Gamma_1(I)$. In this appendix, we state without proof the sum formulas for $\Gamma_1(I)$ resulting from the Theorems 3.21 and 6.5.

Note that we have required $a \equiv d \equiv 1 \mod I$ for $\begin{pmatrix} a & b \\ c & d \end{pmatrix} \in \Gamma_1$, not $a \equiv d \equiv \pm 1 \mod I$. So $m(-1) = \begin{pmatrix} -1 & 0 \\ 0 & -1 \end{pmatrix} \in \Gamma_1$ if and only if $2 \in I$.

We shall use a preceding superscript ι for corresponding objects for Γ_1. For instance, ${}^\iota E_q(\lambda; \nu, i\mu)$ will denote an Eisenstein series for Γ_1, at the cusp λ of $\Gamma_1(I)$. The variable λ runs over a system $\mathcal{P}$ of representatives of the $\Gamma_1(I)$-orbits of cusps. For each λ the μ run through a lattice ${}^\iota\Lambda_\lambda$ in the hyperspace $S(x) = 0$ in $\mathbb{R}^d$.

The Fourier expansion of the Eisenstein series ${}^\iota E_q(\lambda; \nu, i\mu)$ has the same structure as that of $E_q(\kappa, \chi; \nu, i\mu)$ in (2.31), where we now denote the Fourier coefficients by ${}^\iota D(\lambda; \nu, i\mu)$.

Similarly, there are Fourier coefficients ${}^\iota c^r({}^\iota\varpi)$ as in (2.27), where ${}^\iota\varpi$ runs over a maximal orthogonal system of irreducible cuspidal representations in $L^2(\Gamma_1(I)\backslash G)$. The scalar products in $L^2(\Gamma_0(I)\backslash G)$ and in $L^2(\Gamma_1(I)\backslash G)$ are both given by the same Haar measure on G. For $f, g \in L^2(\Gamma_0(I)\backslash G, \chi)$:

(A.1) $$\langle f, g \rangle_{\Gamma_1(I)} = |\Gamma_0(I)/\Gamma_1(I)| \, \langle f, g \rangle_{\Gamma_0(I)}.$$

The Kloosterman sums for $\Gamma_1(I)$ have the following form:

(A.2) $$ {}^\iota S(r', r; c) := \sum_{a \bmod c, \, a \equiv 1 \bmod I}^* e^{2\pi i \mathrm{Tr}_{F/\mathbb{Q}}((ra + r'\tilde{a})/c)},$$

where $a\tilde{a} \equiv 1 \mod (c)$.

With these preparations, we can state the sum formula and the Kloosterman sum formula for $\Gamma_1(I)$:

Theorem A.1. *Let $r, r' \in O' \setminus \{0\}$. Let $\xi \in \{0, 1\}^d$, and suppose that $(-1)^{S(\xi)} = 1$ if $2 \in I$. Put $\varepsilon = \mathrm{sign}\, r$, $\varepsilon' = \mathrm{sign}\, r'$. Define $t_I = 2$ if $2 \in I$ and $t_I = 1$ otherwise. For each $\varphi \in T_\xi^{\varepsilon\varepsilon'}(\tau, a)$ with $\frac{1}{4} < \tau < \frac{1}{2}$, $a > 2$, the following equality holds, with absolute convergence of all integrals and sums:*

(A.3) $$\sum_{{}^\iota\varpi} \overline{{}^\iota c^r({}^\iota\varpi)} \, {}^\iota c^{r'}({}^\iota\varpi) \, \varphi(\nu_{{}^\iota\varpi})$$
$$+ 2 \sum_{\lambda \in \mathcal{P}} {}^\iota c_\lambda \sum_{\mu \in {}^\iota\Lambda_\lambda} \int_0^\infty \overline{{}^\iota D^r(\lambda; iy, i\mu)} \, {}^\iota D^{r'}(\lambda; iy, i\mu) \varphi(iy + i\mu) \, dy$$

$$= \begin{cases} t_I \frac{\sqrt{|D_F|}}{(2\pi)^d} \phi_\xi(m(\zeta)) \int_{Y_\xi} \varphi(v)\, d\mathrm{Pl}_\xi(v) \\ \qquad \text{if } \frac{r}{r'} = \zeta^2 \text{ for some } \zeta \in O^*,\ \zeta \equiv 1 \mod I, \\ 0 \quad \text{otherwise.} \end{cases}$$

$$+ \sum_{c \in I\setminus\{0\}} \frac{{}^tS(r',r;c)}{|N(c)|} \left(\mathrm{B}_\xi^{\varepsilon,\varepsilon'} \varphi \right)\left(\frac{4\pi|rr'|^{1/2}}{c} \right).$$

THEOREM A.2. *Let $r, r' \in O' \setminus \{0\}$, and $\xi \in \{0,1\}^d$. If $2 \in I$, suppose that $(-1)^{S(\xi)} = 1$. Put $\varepsilon = \operatorname{sign} r$, $\varepsilon' = \operatorname{sign} r'$.*

Let $\alpha > \frac{1}{2}$, $\beta > 2$. For all $f \in \mathbf{F}_{\alpha,\beta,\xi}^{(1)}$ (see Definition 6.3) the following equalities hold, with absolute convergence of all integrals and sums:

(A.4) $$\sum_{c \in I\setminus\{0\}} \frac{{}^tS(r',r;c)}{|N(c)|} f\!\left(\frac{4\pi|rr'|^{1/2}}{c} \right)$$

$$= \sum_{{}^t\varpi} \overline{{}^tc^r({}^t\varpi)}\, {}^tc^{r'}({}^t\varpi) \left(\left(\mathrm{B}_\xi^{\varepsilon,\varepsilon'}\right)^{-1} f \right)(v_\varpi)$$

$$+ 2 \sum_{\lambda \in \mathcal{P}} {}^tc_\lambda \sum_{\mu \in {}^t\Lambda_\lambda} \int_0^\infty \overline{{}^tD^r(\lambda; iy, i\mu)}$$

$$\cdot {}^tD^{r'}(\lambda; iy, i\mu) \left(\left(\mathrm{B}_\xi^{\varepsilon,\varepsilon'}\right)^{-1} f \right)(iy + i\mu)\, dy.$$

The ${}^tc_\lambda$ are positive constants. The Bessel transformation $\mathrm{B}_\xi^{\varepsilon,\varepsilon'}$ is given in (3.55)–(3.58), and its inverse $\left(\mathrm{B}_\xi^{\varepsilon,\varepsilon'}\right)^{-1}$ in (6.1) and (5.23).

APPENDIX B

Comparisons

B.1. Trivial central character. Theorem 2.7.1 in [7] gives the sum formula for the case $\Gamma\backslash\mathrm{PSL}_2(\mathbb{R})$, so the central character is $\xi = 0 \in \mathbb{Z}^d$ and the character χ is trivial. On the other hand, a general cofinite discrete subgroup $\Gamma \subset G$ is allowed, and the Fourier coefficients are taken at arbitrary cusps. Let us compare the sum formula in Theorem 3.21 with that in [7], for the situation where there is an overlap. Table 3 gives a comparison of the differences in normalization.

So we have $\Gamma = \Gamma_0(I)$ with I a non-zero ideal in O. We take the cusps κ and κ' in [7] equal to ∞.

The normalization of the Haar measure on N in [7] follows the choice in [35]. In this paper, we preferred to stay close to the usual normalization on $\mathfrak{H}^d$. We have arranged the choice of the factor $d^r(q, v)$ in such a way that the measures on the spectral side of the sum formula coincide.

	Theorem 2.7.1 in [7]		Theorem 3.21 here	
Def. 2.3.1	$\kappa = \kappa' = \infty$		$\xi = 0,\ \chi = 1$	§3.3.1, §2.1
§2.1	dn_{there}	$=$	$\pi^{-d}\, dn_{\text{here}}$	§2.1.2
	dg_{there}	$=$	$\pi^{-d}\, dg_{\text{here}}$	
(15)	$\psi_{\varpi,q,\text{there}}$	$=$	$\pi^{d/2} \psi_{\varpi,q,\text{here}}$	(2.26)
	$\mathrm{vol}_{\text{there}}(\Delta_\infty \backslash N)$	$=$	$\pi^{-d}\sqrt{|D_F|}$	
(16)	$d^r_\infty(q,v)_{\text{there}}$	$=$	$\pi^{d/2} d^r(q,v)_{\text{here}}$	(2.28)
(17)	$c^r_\infty(\varpi)_{\text{there}}$	$=$	$c^r(\varpi)_{\text{here}}$	(2.27)
(12)	$E_q(P^\kappa, v, i\mu)$	$=$	$E_q(\kappa, 1; v, i\mu)$	(2.14)
(18)	$D^{\infty,\infty}_\kappa(v, i\mu)$	$=$	$\pi^{-d/2} D^r_\xi(\kappa, 1; v, i\mu)$	(2.31)
(14)	$c_{\kappa,\text{there}}$	$=$	$\pi^d c_{\kappa,\text{here}}$	(2.20)
Def. 2.5.1	$\mathcal{K}^{\mathbf{e}}$	$\supset$	$T^{\varepsilon,\varepsilon'}_\xi(\tau, a)$	Def. 3.11
(19)	$d\sigma^{\infty,\infty}_{r,r'}$	$=$	$d\sigma^{r,r'}_{1,0}$	(3.43)
	with $\mathbf{e} = \varepsilon\varepsilon'$, $\tfrac{1}{2} < \tau < 1$			
(24)	$B_{\mathbf{e}}\varphi(t)$	$=$	$B^{\varepsilon,\varepsilon'}_0 \varphi(4\pi\sqrt{t})$	(3.55), (3.57)
§2.4	$S(\infty, -r; \infty, -r'; c)$	$=$	$S_1(r', r; c)$	(2.32)
(22), §2.4	$K^{\infty,\infty}_{r,r'}(f)$	$=$	$K^{r,r'}_1(F)$	(3.60)
	with $f(t)$	$=$	$F\left(4\pi\sqrt{t}\right)$	
Def. 2.6.1	$\alpha(\infty, r; \infty, r')_{\text{there}}$	$=$	$\alpha(1, 0; r, r')_{\text{here}}$	(3.48)
Def. 2.5.2	$d\eta^{\mathbf{p}}$	$=$	$2^{-d}\, d\mathrm{Pl}_0$	(3.49)
Def. 2.6.2	$\Delta^{\infty,\infty}_{r,r'}(\varphi)$	$=$	$\dfrac{|D_F|^{1/2} \alpha(1,0;r,r') \int_{Y_\xi} \varphi\, d\mathrm{Pl}_0}{\pi^{-d} 2^{-d}}$	(3.84)

TABLE 3. Relation between the normalizations in [7] and in the present work.

74   B. COMPARISONS

	Theorem 16.4.6 in [4]		Theorem 3.22 here
			$d=1$, $I = N\mathbb{Z}$
	τ_{there}	$=$	$\xi_{\text{here}} \in \{0,1\}$
	σ_{there}	$=$	$\tau_{\text{here}} \in \left(\frac{1}{2},1\right)$
(7.3.7)	$\varepsilon(\alpha)$, $\varepsilon(\beta)$		ε, ε'
	n_α, n_β		r, r'
Def. 14.2.7	$\left({}_{\varepsilon(\alpha)\varepsilon(\beta)}^{1}F^a_{\tau,\sigma}\right)_{\text{there}}$	$\supset$	$\left(T^{\varepsilon\varepsilon'}_\xi(\tau,a)\right)_{\text{here}}$
			Def. 3.10, 3.11
p. 190	$f_{s,j}$	$=$	f_l with $\nu_l = s$ (3.88)
(16.4.22)	$\rho(s,\alpha)_j$	$=$	$\rho_l(r)$ (3.89)
p. 191	$f^1_{b,j}$	$=$	$f^1_{b,l}$ (3.91)
(16.4.27)	$\rho\left(\frac{b-1}{2},\alpha\right)^1_j$	$=$	$\rho^1_{b,l}(r)$ (3.92)
p. 191	$f^{-1}_{b,j}$	$=$	$\overline{f^{-1}_{b,l}}$ (3.95)
(16.4.27)	$\rho\left(\frac{b-1}{2},\alpha\right)^{-1}_j$	$=$	$\overline{\rho^{-1}_{b,l}(r)}$ (3.96)
(7.2.7)	Λ^0	$=$	$\mathcal{P}_\chi$ §2.1.4
(7.2.1)	g^{there}_α	$=$	$\tilde{g}^{\text{here}}_\infty = a(w_\kappa)$ p. 43
(7.2.9)	$\Gamma^{\text{there}}_\infty$	$=$	$\Gamma \cap g_\kappa N g_\kappa^{-1}$ (2.8)
	$\Delta^{\text{there}}_\kappa$	$=$	$\Gamma^{\text{here}}_\kappa$
(9.3.1),	$E\left(\frac{1}{2}+\nu;g_\kappa^{-1},\xi,\chi\right)$	$=$	$w_\kappa^{-\frac{1}{2}-\nu}E_\xi(\kappa,\chi;\nu)$ (2.14)
(9.3.2)	(notation simplified)		
(16.4.15)	$\rho_\kappa(\nu,\alpha)$	$=$	$w_\kappa^{-\frac{1}{2}-\nu}\rho(\kappa,r;\nu)$ (3.97)
	$\Gamma\left(\frac{1}{2}+\nu+\frac{\varepsilon(\alpha)\tau}{2}\right)\Gamma\left(\frac{1}{2}-\nu+\frac{\varepsilon(\beta)\tau}{2}\right)$	$=$	$(\varepsilon\varepsilon')^\xi\Gamma\left(\frac{1}{2}-\nu+\frac{\varepsilon\xi}{2}\right)\Gamma\left(\frac{1}{2}+\nu+\frac{\varepsilon'\xi}{2}\right)$
(14.2.53)	$\langle\varphi,1\rangle$	$=$	$\frac{1}{2}\int_{Y_\xi}\varphi\,d\text{Pl}_\xi$ (3.49)
Def. 8.4.2	$C_{\alpha,\beta}$	$=$	$N\mathbb{N}$
(14.2.11)	$b^{\varepsilon\varepsilon'}_\xi\delta_t(\nu)$	$=$	$2(\varepsilon')^\xi k^{\varepsilon,\varepsilon'}_\xi(\nu,t)$ (3.34)
	for $t > 0$		
(14.2.31)	$\left(b^{\varepsilon\varepsilon'}_\xi\right)^{\leftarrow}\varphi(t)$	$=$	$(-\varepsilon)^\xi B^{\varepsilon,\varepsilon'}_\xi\varphi(t)$
			(3.36), (3.49), (3.53)
Def. 8.4.4	$S(\alpha,\beta;c)$	$=$	$S_\chi(r',r;c)$ $(c>0)$ (2.32)

TABLE 4. Relation between quantities in [4] and the present paper.

In the general setting of Theorem 2.7.1 in [7], there need not be a Weil type estimate as in (2.47), and the variable c in the sum of Kloosterman sums need not run over a lattice. So the strip on which the factors of the test function live has width 2τ, $\frac{1}{2} < \tau < 1$. The parameter **e** in loc. cit. is equal to $\varepsilon\varepsilon' = \text{sign}(rr')$.

The Bessel transforms are related by $f(t) = F\left(4\pi\sqrt{t}\right)$, $t \in (0,\infty)^d$, and $F(y) = |N(y)| f(y^2/16\pi^2)$, $y \in (\mathbb{R}^*)^d$. Note that the Bessel transforms are even functions if $\xi = 0$. This leads to agreement of the Kloosterman terms. More easily, we arrive at equality of the delta terms.

B.2. Comparison with [4], for $d = 1$. Theorem 16.3.7 and Proposition 16.4.6 in [4] give a sum formula for the universal covering group of $SL_2(\mathbb{R})$. The specialization to integral weights should confirm Proposition 3.22. Let us carry out the comparison; see also Table 4.

Proposition 16.4.6 of [**4**] holds for more general cofinite subgroups than the congruence subgroups $\Gamma_0(N)$, so no Weil type estimate of Kloosterman sums can be used there and the test functions have to be holomorphic on a wider strip. So $\tau \in \left(\frac{1}{2}, 1\right)$ in this comparison. We take $\alpha = (\infty, r)$ and $\beta = (\infty, r')$ in Proposition 16.4.6 of [**4**].

The orthonormal systems of cusp forms in Proposition 16.4.6 of [**4**] correspond to the systems used in Proposition 3.22 here, with the exception that [**4**] uses the holomorphic cusp forms $\overline{f_{b,l}^{-1}}$. In [**4**], the cusps are described with $\tilde{g}_\kappa$, so there is a factor $w_\kappa^{-\frac{1}{2}-\nu}$ in the Eisenstein series. With the correspondences given in Table 4, we see that the spectral term in Proposition 16.4.6 in [**4**] is $\pi|rr'|^{1/2}(\varepsilon\varepsilon')^\xi$ times the spectral term in Proposition 3.22.

To obtain the same factor on the geometric side, more comparisons are indicated in Table 4. For the delta term, we also use that $\alpha(\chi, \xi; r, r') = 2\delta_{r,r'}$, and $D_\mathbb{Q} = 1$.

Bibliography

[1] A.Borel, *Introduction aux groupes arithmétiques*, Act. scient. et industr. **1341**; Hermann, Paris, 1969.
[2] A.Borel, *Automorphic forms on* $SL_2(\mathbb{R})$, Cambr. Tracts in Math. **130**, Cambridge University Press, 1997.
[3] R.W.Bruggeman, *Fourier coefficients of cusp forms*, Inv. math. **45** (1978) 1-18.
[4] R.W.Bruggeman, *Fourier Coefficients of Automorphic Forms*, Lecture Notes in Math., **865**, Springer-Verlag, Berlin, 1981.
[5] R.W.Bruggeman, R.J.Miatello, *Estimates of Kloosterman sums for groups of real rank one*, Duke Math. J. **80** (1995) 105–137.
[6] R.W.Bruggeman, R.J.Miatello, *Sum formula for* SL_2 *over a number field and Selberg type estimate for exceptional eigenvalues*, Geom. Funct. Anal. **8** (1998) 627–655.
[7] R.W.Bruggeman, R.J.Miatello, I.Pacharoni, *Estimates for Kloosterman sums for totally real number fields*, J. reine angew. Math. **535** (2001) 103–164.
[8] R.W.Bruggeman, R.J.Miatello, I.Pacharoni, *Sums of Kloosterman sums for real quadratic number fields*, J. Number Th. **99** (2003) 90–119.
[9] R.W.Bruggeman, R.J.Miatello, I.Pacharoni, *Density results for automorphic forms on Hilbert modular groups*, Geometric and Functional Analysis **13** (2003) 681-719.
[10] R.W.Bruggeman, Y.Motohashi, *Sum formula for Koosterman sums and fourth moment of the Dedekind zeta-function over the Gaussian number field*, Functiones et Approximatio **31** (2003) 7–76.
[11] J.W.Cogdell, I.Piatetski-Shapiro, *The Arithmetic and Spectral Analysis of Poincaré Series*, Perspectives in Mathematics **13**, Academic Press, 1990.
[12] J.-M.Deshouillers, H.Iwaniec, *Kloosterman Sums and Fourier Coefficients of Cusp Forms*, Invent. math. **70** (1982), 219–288.
[13] Th.Estermann, *On Kloosterman's sum*, Mathematika **8** (1961) 83-86.
[14] E.Freitag, *Hilbert Modular Forms*, Springer Verlag, 1980.
[15] A.Good, *Local analysis of Selberg's trace formula*, Lect. Notes in Math. **1040**, Springer-Verlag 1984.
[16] K.B.Gundlach, *Über die Darstellung der ganzen Spitzenformen zu den Idealstufen der Hilbertschen Modulgruppe und die Abschätzung ihrer Fourierkoeffizienten*, Acta Math. **92** (1954) 309-345.
[17] Harish Chandra, *Automorphic forms on semisimple Lie groups*, Lect. Notes in Math. **62**, Springer-Verlag, Berlin, 1968.
[18] D.A.Hejhal, *The Selberg trace formula for* $PSL(2,\mathbb{R})$, Lect. Notes in Math. **1001**, Springer-Verlag, 1983.
[19] L.Hörmander, *The Analysis of Linear Partial Differential Operators I*, Grundl. math. Wiss. **256**, Springer, 1983.
[20] A.Hurwitz, *Die unimodularen Substitutionen in einem algebraischen Zahlkörper*, Nachr. k. Gesellsch. Wiss. Göttingen, Math.-phys. Klasse, 1895 (332–356); also in A.Hurwitz, Mathematische Werke, Band II, p. 244–268.
[21] H.Iwaniec, *Introduction to the spectral theory of automorphic forms*, Revista Matemática Iberoamericana, Madrid, 1995.
[22] H.Jacquet, *Smooth transfer of Kloosterman integrals*, Duke Math. J. **120** (2003) 121–152.
[23] H.Jacquet, S.Rallis, *Kloosterman integrals for skew symmetric matrices*, Pac. J. Math. **154** (1992) 265–283.
[24] H.H.Kim, F.Shahidi, *Cuspidality of symmetric powers with applications*, Duke Math. J. **112** (2002) 177–197.
[25] H.H.Kim, F.Shahidi, *Functorial products for* $GL_2 \times GL_3$ *and the symmetric cube for* GL_2, Ann. Math. **155** (2002) 837–893.
[26] H.D.Kloosterman, *On the representation of numbers in the form* $ax^2 + by^2 + cz^2 + dt^2$, Acta Math. **49** (1926) 407–464.
[27] H.D.Kloosterman, *Asymptotische Formeln für die Fourierkoeffizienten ganzer Modulformen*, Abh. Math. Sem. Hamburg **5** (1927) 337–352.

BIBLIOGRAPHY

[28] N.V.Kuznetsov, *The conjecture of Petersson for forms of weight zero and the conjecture of Linnik*, Preprint Khab. KNII. DVNTs. AN.USSR 02-77, Khabarovsk, 1977 (in Russian).

[29] N.V.Kuznetsov, *The Petersson conjecture for parabolic forms of weight zero and the conjecture of Linnik. Sums of Kloosterman sums*, Mat. Sb., **111** (1980), 334–383 (in Russian).

[30] S.Lang, $SL_2(\mathbb{R})$, Addison-Wesley, 1975.

[31] R.P.Langlands, *On the functional equations satisfied by Eisenstein series*, Lect. Notes in Math. **544**, Springer-Verlag, Berlin, 1976.

[32] N.N.Lebedev, *Special functions and their applications*, Dover Publ., 1972.

[33] H.Lokvenec-Guleska, *Sum formula for SL_2 over an Imaginary Quadratic Number Field*, thesis, Utrecht, Nov. 2004;
http://igitur-archive.library.uu.nl/dissertations/2004-1203-105121/index.htm.

[34] R.J.Miatello, N.R.Wallach, *Automorphic forms constructed from Whittaker vectors*, J. Funct. Anal. **86** (1989) 411-487.

[35] R.J.Miatello, N.R.Wallach, *Kuznetsov formulas for real rank one groups*, J. Funct. Anal. **93** (1990) 171–206.

[36] G.D.Mostow, *Discrete subgroups of Lie groups*, p. 65–153 in *Lie theories and their applications*, Queen's papers in pure and applied mathematica **48**, Kingston, 1978.

[37] Y.Motohashi, *Spectral theory of the Riemann zeta-function*, Cambridge Tracts in Mathematics **127**, Cambridge University Press, 1997.

[38] Y.Motohashi, *A note on the mean value of the zeta and L-functions. XII*, Proc. Japan Acad., **78A** (2002), 36-41.

[39] D.Niebur, *A class of nonanalytic automorphic functions*, Nagoya Math. J. **52** (1973) 133–145.

[40] H.Petersson, *Über die Entwicklungskoeffizienten der automorphen Formen*, Acta Math. **58** (1932) 169–215.

[41] N.V.Proskurin, *Sum formulas for general Kloosterman sums*, Zap. nauchn. seminarov LOMI **82** (1979) 103–135 (in Russian).

[42] H.Salié, *Über die Kloostermanschen Summen $S(u, v; q)$*, Math. Z. **34** (1931) 91–109.

[43] A.Selberg, *On the estimation of Fourier coefficients of modular forms*, Proc. SPM VIII, AMS, 1965, 1–15.

[44] L.J.Slater, *Confluent hypergeometric functions*, Cambridge, at the University Press, 1960.

[45] C.M.Sorensen, *Fourier expansion of Eisenstein series on the Hilbert modular group and Hilbert class fields*, Trans. AMS **354** (2002) 4847–4869.

[46] E.C.Titchmarsh, *The theory of the Riemann zeta-function*, Oxford, at the Clarendon Press, 1956.

[47] A.Venkatesh, *Beyond endoscopy and special forms on* GL(2), J. reine angew. Math. **577** (2004) 23–80.

[48] A.B.Venkov, *Spectral theory of automorphic functions and its applications*, Kluwer, 1990.

[49] N.R.Wallach, *Real Reductive Groups. II*, Pure and Applied Mathematics, **132**, Academic Press, 1992.

[50] G.N.Watson, *A treatise on the theory of Bessel functions*, second edition, Cambridge, at the University Press, 1944.

[51] N.R.Wallach, R.Miatello, *Kuznetsov formulas for Products of Groups of $\mathbb{R}$-rank One*, Israel Math. Conf. Proceedings **3** (volume in honor of I.Piatetskii-Shapiro), 1990, 305-321.

[52] A.Weil, *On some exponential sums*, Proc. Nat. Acad. Sci. U.S.A. **34** (1948) 204–207.

[53] Y.Ye, *Kloosterman integrals and base change for* GL(2), J. reine angew. Math. **400** (1989) 57–121.

[54] Y.Ye, *An integral transform and its applications*, Math. Ann. **300** (1994) 405–417.

Index

antiholomorphic discrete series   10, 44
automorphic form   7
automorphic representation   11
automorphic transformation rule   5, 7

Bessel function   22, 32, 61–64
Bessel inversion   58
Bessel transform   4, 28, 31, 33
big cell   17, 23, 28
Bruhat decomposition   17

Casimir operator   7
center   6
central character   11
classification of representations of $SL_2(\mathbb{R})$ 10
complementary ideal   8
complementary series   10, 44
continuous spectrum   26
convolution   23
cusp   6
cusp form   9

delta term in sum formula   4, 27, 59
different   13
discrete series   10
discrete spectrum   26
discriminant   8

Eisenstein series   3, 8
exceptional eigenvalue   39
exceptional spectral parameter   39

K-finite   6
Fourier coefficient   12
Fourier expansion   3, 7
Fourier term   8
fundamental domain   9

geometric side of sum formula   4, 27
Götzky-Koecher principle   8
growth condition   7

Haar measure   6
Hilbert modular group   5
holomorphic automorphic forms   7
holomorphic discrete series   10, 11, 44

invariant measure   6
inverse Bessel transform   4, 55, 60
Iwasawa coordinates   6
Iwasawa decomposition   6

Kloosterman integrals   70
Kloosterman sum   4, 13, 17
Kloosterman sum formula   67, 72
Kloosterman term in sum formula   30

Lie algebra   10
local Bessel transform   29, 32
local test functions   31

Maass form   3
Mellin transform   56
mock discrete series   10
modular group   46

orthonormal basis   9, 12

Plancherel measure   28
Poincaré series   15, 25
principal congruence subgroup   52

Ramanujan conjecture   39
real analytic   7
relative trace formula   70
restricted sum formula   30
right inverse   58, 60
right translation   9

spectral decomposition   9
spectral parameter   7, 10
spectral side of sum formula   4, 26
sum formula   3, 41, 45, 71
sum of Kloosterman sums   30, 33
sum over the units   15

test function   3, 31

unitary principal series   10, 44
universal enveloping algebra   10

weight   6
weight subspace   9
Weil type estimate   13

Whittaker transform   19

Notations

(A1) – (A4) : conditions   7
A : $\{a(y)\}$ 6
da : Haar measure on A 6
$a(y)$: matrix $\begin{pmatrix} \sqrt{y} & 0 \\ 0 & 1/\sqrt{y} \end{pmatrix}$ (2.3)

$\mathcal{B}^{\pm 1}$, $\mathcal{B}_j^{\pm 1}$: Bessel differential operator   (6.30)
$B = B_\xi^{\varepsilon,\varepsilon'}$: Bessel transform   (3.53), (3.54)
$B^* = \left(B_\xi^{\varepsilon,\varepsilon'}\right)^*$: adjoint of Bessel transform (6.1)

C_j : Casimir operator   7
C_ν (3.28)
c_κ, $^tc_\lambda$: constant in spectral decomposition (2.20), (3.86), 72
$c^r(\varpi)$, $^tc^r(\varpi)$: Fourier coefficient   (2.27), 71
$C^r(X)$ sum of eigenvalue vectors : (4.3)

$D(\xi,\tau)$: domain for the test functions   18
D_F : discriminant   8
D_j : $t_j \partial_{t_j}$ 65
$D_\xi^r(\kappa,\chi;\nu,i\mu)$, $^tD_\xi^r(\lambda;\nu,i\mu)$: Fourier coefficient of Eisenstein series   (2.31), 71
$d^r(q,\nu)$: normalization factor in Fourier term (2.28)

$\mathbf{E}_j^\pm$: weight shifting operator   (2.21)
$E_q(\kappa,\chi;\nu,i\mu)$, $^tE_q(\lambda;\nu,i\mu)$: Eisenstein series (2.14), 71

$\mathbf{F}_{\alpha,\beta}$, $\mathbf{F}_{\alpha,\beta}^{(1)}$: spaces of functions on $(\mathbb{R}^*)^d$ 61, 65
$\mathbf{F}_{\alpha,\beta,\xi}$: subspace of $\mathbf{F}^{\alpha,\beta}$ determined by parity condition   61
F : totally real number field   5
F_r : Fourier term operator   (2.11)
$\mathcal{F}_G$, $\mathcal{F}_{\tilde{\mathfrak{z}}}$, $\mathcal{F}_K$: fundamental domains   9
f_l, $f_{b,l}^{\pm 1}$: automorphic forms   44–45

G : $SL_2(\mathbb{R})^d$ 5
$\mathfrak{g}$: Lie algebra   10
dg : Haar measure on G 6, 23
g_κ : $\kappa = g_\kappa \cdot \infty$ 6
$\tilde{g}_\kappa$: $g_\kappa a(w_\kappa)$ 43

$H_{\delta,\tau}$, $H_{q,\tau}^r$: spaces of auxiliary test functions 18

I : ideal in O 5
$I^0(\eta,\eta',u;\nu)$, $I^\infty(\eta,\eta',u;\nu)$: integrals of Bessel function   (6.5)

$J_u^1 = J_u$, $J_u^{-1} = I_u$: Bessel functions   (3.27)
$j^\pm$: Bessel transform   (5.3)

K : $SO_2(\mathbb{R})^d$ 6
$K_\chi^{r,r'}(\cdot)$: sum of Kloosterman sums   (3.60)

$k_\xi^{\varepsilon,\varepsilon'}$: Bessel kernel   (3.35)
dk : Haar measure on K 6
$k(\vartheta)$: matrix $\begin{pmatrix} \cos\vartheta & \sin\vartheta \\ -\sin\vartheta & \cos\vartheta \end{pmatrix}$ (2.3)

$L^2(\Gamma\backslash G,\chi)$ 8
$L_\xi^2(\Gamma\backslash G,\chi)$: subspace with central character specified by ξ 11
$L^2(\Gamma\backslash G,\chi)_q$: weight subspace   9
$L^{2,\text{discr}}(\Gamma\backslash G,\chi)_q$, $L^{2,\text{cont}}(\Gamma\backslash G,\chi)_q$ 9

$\mathcal{M}$: Mellin transform   (5.5)
M : center of G 6
M_ν^r : local Whittaker function   (3.20)
$m(\zeta)$: element of M (2.4)
$m(\tau,t)$: $\prod_j \min\left(|t_j|^{2\tau},1\right)$ 34

$\mathbb{N}_0$: $\mathbb{Z}_{\geq 0}$ 5
N : unipotent subgroup   6
$N(\cdot)$: norm extending $N_{F/\mathbb{Q}}$ 5
$N_{\alpha,\beta}$, $N_{\alpha,\beta}^{(1)}$: norms on functions on $(\mathbb{R}^*)^d$ (6.4), (6.40)
$n(\cdot,\cdot)$: normalization factor   (2.26)
dn : Haar measure on N 6
$n(x)$: matrix $\begin{pmatrix} 1 & x \\ 0 & 1 \end{pmatrix}$ (2.3)

O : integers in F 5
O' : complementary ideal   8

$\mathcal{P}$: representatives of cuspidal orbits   6
$^t\mathcal{P}$: representatives of cuspidal orbits for $\Gamma_1(I)$ 71
$\mathcal{P}_\chi$: subset of $\mathcal{P}$ 6
Ph : Poincaré series with seed function h (2.48)
$d\text{Pl}_\xi$: Plancherel measure   28

R_g : right translation   9

$S(\cdot)$: trace   5
$S_\chi(r,r';c)$: Kloosterman sum   (2.32)
$^tS(r',r;c)$: Kloosterman sum for $\Gamma_1(I)$ (A.2)

$T_\xi^{\varepsilon\varepsilon'}(\tau,a)$: principal test functions   31

$\mathcal{U}(\mathfrak{g})$: universal enveloping algebra   10

V_ν (3.23)

W_q : product of Whittaker functions   (2.12)
W_ν^r (3.26)
w_κ : width of cusp κ 43
w_q^r : Whittaker transform   19
w_u : local Whittaker transform   (3.10)

Y_ξ : spectral parameter space   (3.42)

$\alpha(\chi,\xi;r,r')$: factor in delta term   (3.48)

$\Gamma_0(I)$: congruence subgroup   5
Γ : $\Gamma_0(I)$ 5
$\Gamma_1(I) = \Gamma_1$: congruence subgroup   5, §A

INDEX

$\Gamma(I)$: principal congruence subgroup 52
Γ_κ : isotropy subgroup of κ in Γ 6
Γ_N : $\Gamma_\infty \cap N$ (2.8)

ε : sign r 25

$\vartheta_q^{r,r'}$ (3.45)

$\Lambda_{\kappa,\chi}$, ${}^t\Lambda_\lambda$: lattices in hyperplane $S(x) = 0$ 8, 71, 71

$d\mu$: invariant measure   (2.5)

ξ : parameter determining central character   11, 25

σ_j : embedding $F \to \mathbb{R}$ 5
$d\sigma_{\chi,\xi}^{r,r'}$: measure describing spectral side of sum formula   (3.43)

ϕ_q : character of K 6

χ : character of $(O/I)^*$ 5
χ_r : character of N 7

$\psi_{\varpi,q}$: basis element   12

ω_u : local Whittaker transform   (3.10)

$\|\cdot\|$: norm in $L^2(\Gamma\backslash G, \chi)$ 8
$\|\mu\|$: norm in lattice   52
$\langle\cdot,\cdot\rangle$: scalar product in $L^2(\Gamma_0(I)\backslash G, \chi)$ 8
$\langle\cdot,\cdot\rangle_{\Gamma_1}$: scalar product in $L^2(\Gamma_1(I)\backslash G)$ (A.1)
$\|\cdot\|_{\Gamma\backslash\mathfrak{H}}$, $\langle\cdot,\cdot\rangle_{\Gamma\backslash\mathfrak{H}}$: norm and scalar product in $\Gamma\backslash\mathfrak{H}$ (3.85)

Editorial Information

To be published in the *Memoirs*, a paper must be correct, new, nontrivial, and significant. Further, it must be well written and of interest to a substantial number of mathematicians. Piecemeal results, such as an inconclusive step toward an unproved major theorem or a minor variation on a known result, are in general not acceptable for publication.

Papers appearing in *Memoirs* are generally at least 80 and not more than 200 published pages in length. Papers less than 80 or more than 200 published pages require the approval of the Managing Editor of the Transactions/Memoirs Editorial Board.

As of September 30, 2008, the backlog for this journal was approximately 15 volumes. This estimate is the result of dividing the number of manuscripts for this journal in the Providence office that have not yet gone to the printer on the above date by the average number of monographs per volume over the previous twelve months, reduced by the number of volumes published in four months (the time necessary for preparing a volume for the printer). (There are 6 volumes per year, each usually containing at least 4 numbers.)

A Consent to Publish and Copyright Agreement is required before a paper will be published in the *Memoirs*. After a paper is accepted for publication, the Providence office will send a Consent to Publish and Copyright Agreement to all authors of the paper. By submitting a paper to the *Memoirs*, authors certify that the results have not been submitted to nor are they under consideration for publication by another journal, conference proceedings, or similar publication.

Information for Authors

Memoirs are printed from camera copy fully prepared by the author. This means that the finished book will look exactly like the copy submitted.

Initial submission. The AMS uses Centralized Manuscript Processing for initial submissions. Authors should submit a PDF file using the Initial Manuscript Submission form found at www.ams.org/peer-review-submission, or send one copy of the manuscript to the following address: Centralized Manuscript Processing, MEMOIRS OF THE AMS, 201 Charles Street, Providence, RI 02904-2294 USA. If a paper copy is being forwarded to the AMS, indicate that it is for it Memoirs and include the name of the corresponding author, contact information such as email address or mailing address, and the name of an appropriate Editor to review the paper (see the list of Editors below).

The paper must contain a *descriptive title* and an *abstract* that summarizes the article in language suitable for workers in the general field (algebra, analysis, etc.). The *descriptive title* should be short, but informative; useless or vague phrases such as "some remarks about" or "concerning" should be avoided. The *abstract* should be at least one complete sentence, and at most 300 words. Included with the footnotes to the paper should be the 2000 *Mathematics Subject Classification* representing the primary and secondary subjects of the article. The classifications are accessible from www.ams.org/msc/. The list of classifications is also available in print starting with the 1999 annual index of *Mathematical Reviews*. The Mathematics Subject Classification footnote may be followed by a list of *key words and phrases* describing the subject matter of the article and taken from it. Journal abbreviations used in bibliographies are listed in the latest *Mathematical Reviews* annual index. The series abbreviations are also accessible from www.ams.org/msnhtml/serials.pdf. To help in preparing and verifying references, the AMS offers MR Lookup, a Reference Tool for Linking, at www.ams.org/mrlookup/.

Electronically prepared manuscripts. The AMS encourages electronically prepared manuscripts, with a strong preference for $\mathcal{A}_{\mathcal{M}}\mathcal{S}$-LaTeX. To this end, the Society has prepared $\mathcal{A}_{\mathcal{M}}\mathcal{S}$-LaTeX author packages for each AMS publication. Author packages include instructions for preparing electronic manuscripts, samples, and a style file that generates the particular design specifications of that publication series. Though $\mathcal{AMS}$-LaTeX is the highly preferred format of TeX, author packages are also available in $\mathcal{AMS}$-TeX.

Authors may retrieve an author package for *Memoirs of the AMS* from www.ams.org/journals/memo/memoauthorpac.html or via FTP to ftp.ams.org (login as anonymous, enter username as password, and type cd pub/author-info). The *AMS Author Handbook* and the *Instruction Manual* are available in PDF format from the author package link. The author package can also be obtained free of charge by sending email to tech-support@ams.org (Internet) or from the Publication Division, American Mathematical Society, 201 Charles St., Providence, RI 02904-2294, USA. When requesting an author package, please specify $\mathcal{AMS}$-LaTeX or $\mathcal{AMS}$-TeX and the publication in which your paper will appear. Please be sure to include your complete mailing address.

After acceptance. The final version of the electronic file should be sent to the Providence office (this includes any TeX source file, any graphics files, and the DVI or PostScript file) immediately after the paper has been accepted for publication.

Before sending the source file, be sure you have proofread your paper carefully. The files you send must be the EXACT files used to generate the proof copy that was accepted for publication. For all publications, authors are required to send a printed copy of their paper, which exactly matches the copy approved for publication, along with any graphics that will appear in the paper.

Accepted electronically prepared files can be submitted via the web at www.ams.org/submit-book-journal/, sent via FTP, or sent on CD-Rom or diskette to the Electronic Prepress Department, American Mathematical Society, 201 Charles Street, Providence, RI 02904-2294 USA. TeX source files, DVI files, and PostScript files can be transferred over the Internet by FTP to the Internet node ftp.ams.org (130.44.1.100). When sending a manuscript electronically via CD-Rom or diskette, please be sure to include a message identifying the paper as a Memoir.

Electronically prepared manuscripts can also be sent via email to pub-submit@ams.org (Internet). In order to send files via email, they must be encoded properly. (DVI files are binary and PostScript files tend to be very large.)

Electronic graphics. Comprehensive instructions on preparing graphics are available at www.ams.org/authors/journals.html. A few of the major requirements are given here.

Submit files for graphics as EPS (Encapsulated PostScript) files. This includes graphics originated via a graphics application as well as scanned photographs or other computer-generated images. If this is not possible, TIFF files are acceptable as long as they can be opened in Adobe Photoshop or Illustrator. No matter what method was used to produce the graphic, it is necessary to provide a paper copy to the AMS.

Authors using graphics packages for the creation of electronic art should also avoid the use of any lines thinner than 0.5 points in width. Many graphics packages allow the user to specify a "hairline" for a very thin line. Hairlines often look acceptable when proofed on a typical laser printer. However, when produced on a high-resolution laser imagesetter, hairlines become nearly invisible and will be lost entirely in the final printing process.

Screens should be set to values between 15% and 85%. Screens which fall outside of this range are too light or too dark to print correctly. Variations of screens within a graphic should be no less than 10%.

Inquiries. Any inquiries concerning a paper that has been accepted for publication should be sent to memo-query@ams.org or directly to the Electronic Prepress Department, American Mathematical Society, 201 Charles St., Providence, RI 02904-2294 USA.

Editors

This journal is designed particularly for long research papers, normally at least 80 pages in length, and groups of cognate papers in pure and applied mathematics. Papers intended for publication in the *Memoirs* should be addressed to one of the following editors. The AMS uses Centralized Manuscript Processing for initial submissions to AMS journals. Authors should follow instructions listed on the Initial Submission page found at www.ams.org/memo/memosubmit.html.

Algebra to ALEXANDER KLESHCHEV, Department of Mathematics, University of Oregon, Eugene, OR 97403-1222; email: ams@noether.uoregon.edu

Algebraic geometry and its application to MINA TEICHER, Emmy Noether Research Institute for Mathematics, Bar-Ilan University, Ramat-Gan 52900, Israel; email: teicher@macs.biu.ac.il

Algebraic geometry to DAN ABRAMOVICH, Department of Mathematics, Brown University, Box 1917, Providence, RI 02912; email: amsedit@math.brown.edu

Algebraic topology to ALEJANDRO ADEM, Department of Mathematics, University of British Columbia, Room 121, 1984 Mathematics Road, Vancouver, British Columbia, Canada V6T 1Z2; email: adem@math.ubc.ca

Combinatorics to JOHN R. STEMBRIDGE, Department of Mathematics, University of Michigan, Ann Arbor, Michigan 48109-1109; email: FRS@umich.edu

Complex analysis and harmonic analysis to ALEXANDER NAGEL, Department of Mathematics, University of Wisconsin, 480 Lincoln Drive, Madison, WI 53706-1313; email: nagel@math.wisc.edu

Differential geometry and global analysis to LISA C. JEFFREY, Department of Mathematics, University of Toronto, 100 St. George St., Toronto, ON Canada M5S 3G3; email: jeffrey@math.toronto.edu

Dynamical systems and ergodic theory and complex anaysis to YUNPING JIANG, Department of Mathematics, CUNY Queens College and Graduate Center, 65-30 Kissena Blvd., Flushing, NY 11367; email: Yunping.Jiang@qc.cuny.edu

Functional analysis and operator algebras to DIMITRI SHLYAKHTENKO, Department of Mathematics, University of California, Los Angeles, CA 90095; email: shlyakht@math.ucla.edu

Geometric analysis to WILLIAM P. MINICOZZI II, Department of Mathematics, Johns Hopkins University, 3400 N. Charles St., Baltimore, MD 21218; email: trans@math.jhu.edu

Geometric analysis to MARK FEIGHN, Math Department, Rutgers University, Newark, NJ 07102; email: feighn@andromeda.rutgers.edu

Harmonic analysis, representation theory, and Lie theory to ROBERT J. STANTON, Department of Mathematics, The Ohio State University, 231 West 18th Avenue, Columbus, OH 43210-1174; email: stanton@math.ohio-state.edu

Logic to STEFFEN LEMPP, Department of Mathematics, University of Wisconsin, 480 Lincoln Drive, Madison, Wisconsin 53706-1388; email: lempp@math.wisc.edu

Number theory to JONATHAN ROGAWSKI, Department of Mathematics, University of California, Los Angeles, CA 90095; email: jonr@math.ucla.edu

Partial differential equations to GUSTAVO PONCE, Department of Mathematics, South Hall, Room 6607, University of California, Santa Barbara, CA 93106; email: ponce@math.ucsb.edu

Partial differential equations and dynamical systems to PETER POLACIK, School of Mathematics, University of Minnesota, Minneapolis, MN 55455; email: polacik@math.umn.edu

Probability and statistics to RICHARD BASS, Department of Mathematics, University of Connecticut, Storrs, CT 06269-3009; email: bass@math.uconn.edu

Real analysis and partial differential equations to DANIEL TATARU, Department of Mathematics, University of California, Berkeley, Berkeley, CA 94720; email: tataru@math.berkeley.edu

All other communications to the editors should be addressed to the Managing Editor, ROBERT GURALNICK, Department of Mathematics, University of Southern California, Los Angeles, CA 90089-1113; email: guralnic@math.usc.edu.

Titles in This Series

923 **Michael Jöllenbeck and Volkmar Welker,** Minimal resolutions via algebraic discrete Morse theory, 2009

922 **Ph. Barbe and W. P. McCormick,** Asymptotic expansions for infinite weighted convolutions of heavy tail distributions and applications, 2009

921 **Thomas Lehmkuhl,** Compactification of the Drinfeld modular surfaces, 2009

920 **Georgia Benkart, Thomas Gregory, and Alexander Premet,** The recognition theorem for graded Lie algebras in prime characteristic, 2009

919 **Roelof W. Bruggeman and Roberto J. Miatello,** Sum formula for SL_2 over a totally real number field, 2009

918 **Jonathan Brundan and Alexander Kleshchev,** Representations of shifted Yangians and finite W-algebras, 2008

917 **Salah-Eldin A. Mohammed, Tusheng Zhang, and Huaizhong Zhao,** The stable manifold theorem for semilinear stochastic evolution equations and stochastic partial differential equations, 2008

916 **Yoshikata Kida,** The mapping class group from the viewpoint of measure equivalence theory, 2008

915 **Sergiu Aizicovici, Nikolaos S. Papageorgiou, and Vasile Staicu,** Degree theory for operators of monotone type and nonlinear elliptic equations with inequality constraints, 2008

914 **E. Shargorodsky and J. F. Toland,** Bernoulli free-boundary problems, 2008

913 **Ethan Akin, Joseph Auslander, and Eli Glasner,** The topological dynamics of Ellis actions, 2008

912 **Igor Chueshov and Irena Lasiecka,** Long-time behavior of second order evolution equations with nonlinear damping, 2008

911 **John Locker,** Eigenvalues and completeness for regular and simply irregular two-point differential operators, 2008

910 **Joel Friedman,** A proof of Alon's second eigenvalue conjecture and related problems, 2008

909 **Cameron McA. Gordon and Ying-Qing Wu,** Toroidal Dehn fillings on hyperbolic 3-manifolds, 2008

908 **J.-L. Waldspurger,** L'endoscopie tordue n'est pas si tordue, 2008

907 **Yuanhua Wang and Fei Xu,** Spinor genera in characteristic 2, 2008

906 **Raphaël S. Ponge,** Heisenberg calculus and spectral theory of hypoelliptic operators on Heisenberg manifolds, 2008

905 **Dominic Verity,** Complicial sets characterising the simplicial nerves of strict ω-categories, 2008

904 **William M. Goldman and Eugene Z. Xia,** Rank one Higgs bundles and representations of fundamental groups of Riemann surfaces, 2008

903 **Gail Letzter,** Invariant differential operators for quantum symmetric spaces, 2008

902 **Bertrand Toën and Gabriele Vezzosi,** Homotopical algebraic geometry II: Geometric stacks and applications, 2008

901 **Ron Donagi and Tony Pantev (with an appendix by Dmitry Arinkin),** Torus fibrations, gerbes, and duality, 2008

900 **Wolfgang Bertram,** Differential geometry, Lie groups and symmetric spaces over general base fields and rings, 2008

899 **Piotr Hajłasz, Tadeusz Iwaniec, Jan Malý, and Jani Onninen,** Weakly differentiable mappings between manifolds, 2008

898 **John Rognes,** Galois extensions of structured ring spectra/Stably dualizable groups, 2008

897 **Michael I. Ganzburg,** Limit theorems of polynomial approximation with exponential weights, 2008

TITLES IN THIS SERIES

896 **Michael Kapovich, Bernhard Leeb, and John J. Millson,** The generalized triangle inequalities in symmetric spaces and buildings with applications to algebra, 2008
895 **Steffen Roch,** Finite sections of band-dominated operators, 2008
894 **Martin Dindoš,** Hardy spaces and potential theory on C^1 domains in Riemannian manifolds, 2008
893 **Tadeusz Iwaniec and Gaven Martin,** The Beltrami Equation, 2008
892 **Jim Agler, John Harland, and Benjamin J. Raphael,** Classical function theory, operator dilation theory, and machine computation on multiply-connected domains, 2008
891 **John H. Hubbard and Peter Papadopol,** Newton's method applied to two quadratic equations in $\mathbb{C}^2$ viewed as a global dynamical system, 2008
890 **Steven Dale Cutkosky,** Toroidalization of dominant morphisms of 3-folds, 2007
889 **Michael Sever,** Distribution solutions of nonlinear systems of conservation laws, 2007
888 **Roger Chalkley,** Basic global relative invariants for nonlinear differential equations, 2007
887 **Charlotte Wahl,** Noncommutative Maslov index and eta-forms, 2007
886 **Robert M. Guralnick and John Shareshian,** Symmetric and alternating groups as monodromy groups of Riemann surfaces I: Generic covers and covers with many branch points, 2007
885 **Jae Choon Cha,** The structure of the rational concordance group of knots, 2007
884 **Dan Haran, Moshe Jarden, and Florian Pop,** Projective group structures as absolute Galois structures with block approximation, 2007
883 **Apostolos Beligiannis and Idun Reiten,** Homological and homotopical aspects of torsion theories, 2007
882 **Lars Inge Hedberg and Yuri Netrusov,** An axiomatic approach to function spaces, spectral synthesis and Luzin approximation, 2007
881 **Tao Mei,** Operator valued Hardy spaces, 2007
880 **Bruce C. Berndt, Geumlan Choi, Youn-Seo Choi, Heekyoung Hahn, Boon Pin Yeap, Ae Ja Yee, Hamza Yesilyurt, and Jinhee Yi,** Ramanujan's forty identities for Rogers-Ramanujan functions, 2007
879 **O. García-Prada, P. B. Gothen, and V. Muñoz,** Betti numbers of the moduli space of rank 3 parabolic Higgs bundles, 2007
878 **Alessandra Celletti and Luigi Chierchia,** KAM stability and celestial mechanics, 2007
877 **María J. Carro, José A. Raposo, and Javier Soria,** Recent developments in the theory of Lorentz spaces and weighted inequalities, 2007
876 **Gabriel Debs and Jean Saint Raymond,** Borel liftings of Borel sets: Some decidable and undecidable statements, 2007
875 **C. Krattenthaler and T. Rivoal,** Hypergéométrie et fonction zêta de Riemann, 2007
874 **Sonia Natale,** Semisolvability of semisimple Hopf algebras of low dimension, 2007
873 **A. J. Duncan,** Exponential genus problems in one-relator products of groups, 2007
872 **Anthony V. Geramita, Tadahito Harima, Juan C. Migliore, and Yong Su Shin,** The Hilbert function of a level algebra, 2007
871 **Pascal Auscher,** On necessary and sufficient conditions for L^p-estimates of Riesz transforms associated to elliptic operators on $\mathbb{R}^n$ and related estimates, 2007
870 **Takuro Mochizuki,** Asymptotic behaviour of tame harmonic bundles and an application to pure twistor D-modules, Part 2, 2007
869 **Takuro Mochizuki,** Asymptotic behaviour of tame harmonic bundles and an application to pure twistor D-modules, Part 1, 2007

For a complete list of titles in this series, visit the
AMS Bookstore at **www.ams.org/bookstore/**.